The Last Sunset in the West

Natalie Sanders has a PhD in marine biology and has collaborated with the Hebridean Whale and Dolphin Trust and charities in British Columbia, Canada. She lives in south-west England with her husband and two children.

The Last Sunset in the West

Britain's Vanishing West Coast Orcas

NATALIE SANDERS

BIRLINN

This revised edition first published in Great Britain in 2024 by
Birlinn Ltd
West Newington House
10 Newington Road
Edinbugh
EH9 1QS

www.birlinn.co.uk

First published in hardback in 2023 by Sandstone Press Ltd

ISBN: 978 1 78027 894 0

Typeset by Hewer Text UK Ltd, Edinburgh

Papers used by Birlinn Ltd are from well-managed
forests and other responsible sources

Printed and bound by Clays Ltd, Elcograf S.p.A.

To Comet and Aquarius,
Quinn and Owen

One way to open your eyes to unnoticed beauty is
to ask yourself, 'What if I had never seen this before?
What if I knew I would never see it again?'

RACHEL CARSON, *The Sense of Wonder*
Marine conservationist
(1907–1964)

Contents

Foreword

IN THE LAST SUNSET IN THE WEST, Natalie Sanders offers a contribution to the growing local and personal literature sharing human experiences with whales, dolphins and porpoises. That this book invites the reader into the world of a totemic* species *Orcinus orca,* or simply orca, makes it even more appealing, especially to orca lovers everywhere, or 'orcaholics' as we're sometimes called.

Forty years ago, there were no books on orcas when I published *The Whale Called Killer* (which became *Orca: The Whale Called Killer* in later editions). Now there are two or three dozen orca books, each of them particular to one coastline or another, with close-up personal experiences of individually named orcas feeding, playing, engaging in every aspect of family life, swimming proudly down one strait or fjord or open ocean like they own the place.

What we discover from the best of these books, including *The Last Sunset in the West,* is that the orca story is full of contradictions. On the one hand orcas are the top predator, strong, fast, at times voracious hunters, with nothing that can challenge them in their natural habitat. And on the other hand, they can be vulnerable, as Natalie describes with John Coe and the Scottish orcas in the West Coast Community.

Orcas have a reproductive biology with a very low birth rate and cultural systems that have contributed to the division of the

---

* In Native American culture, a totem animal is the spiritual symbol of an individual, family or tribe. Your totemic animal, according to First Nations' belief, is the main guiding spirit that stays with you for your lifetime and is carried down through your family lineage.

species into at least ten ecotypes which may well one day be declared a number of separate species. Each of these ecotypes are further divided into small, separate breeding populations or communities. This means that if you are an orca mother and your allotted five offspring over your lifetime are all or mostly males, your line is going to die out. If you are an orca community with only a few pods and you get captured by aquarium captors (as happened with the southern orca community off Vancouver Island in the 1960s–70s), or if you run into whalers (about 250 orcas were killed in the Shetlands, 1955–64), that can mark the beginning of the end of an entire breeding population.

Small-boat whaling may indeed explain what happened to the presumed communities of orcas related to John Coe and friends, who are now the last of their kind. The others were simply eliminated, deprived of their rightful places in the seas around us, and never mind the deprivation of Britain losing some of its most iconic, or even totemic, inhabitants. Of course, there can be additional complicating factors leading to an orca community's decline if it is carrying high contaminant loads, or if it is susceptible to fishing interactions that may lead to entanglement, or if it is simply past breeding age, older orcas swimming around and staying together but no longer able to pass their genes to the next generation.

It is not an easy life in the ocean, even for the world's top marine predator. And we as humans share those difficulties as authors of the planet's climate emergency, the biodiversity crisis, the acidification of the ocean and COVID-19. Thus, we can perhaps identify at some level with orca communities who are threatened, indeed in the case of the West Coast Community poised for extinction of their ancestral line. Of course, ultimately, we – every individual of every species on this planet – are all in this together. Humans have been slow to recognise this, but more and more that central fact is impacting our lives.

Natalie's book is as wide-ranging as orcas are on their daily travels, with plenty of science and stories of individual orcas who matter to the people who share their coastlines. *The Last Sunset in the West* offers a comforting tribute to Britain's last orca pod, but it's also a rallying cry that we all need to do more if we want to keep alive the magic of wild orcas and other species in the ocean – Planet Ocean, our shared ocean.

ERICH HOYT

Introduction

THE ORCA IS THE OCEAN'S APEX PREDATOR and most sentient inhabitant, an animal that has since time immemorial fascinated, scared, intrigued and inspired humankind. It has been given many names, but the most common are orca and killer whale. As the largest member of the dolphin family, orcas are not actually whales at all, but it is commonplace to refer to them as such. Regional names are also used such as the Gaelic *mada-chuain*, which is how they are known in the Hebrides.

'Blackfish' and 'kakawin' are commonly used in the Pacific Northwest by local and indigenous cultures. While researchers, who love and respect them, most commonly use 'killer whale', this is increasingly viewed as pejorative and there is a public desire to move away from it. The term 'killer whale' came from Spanish sailors who saw them hunting larger whales in groups. They called them *asesina ballenas,* meaning 'whale killers' but, somewhere along the line, noun and adjective were switched to become 'killer whale'.

Their scientific, Latin, name is *Orcinus orca,* where *Orcinus* translates as 'The kingdom of the dead' and *orca* as 'a kind of whale', which conjures up unfortunate images of them as deadly ferocious, heartless creatures who kill for pleasure. Having spent many years learning about them, observing them in the wild and analysing countless hours of drone footage, there is no denying their hunting prowess, but they are so much more complex. I have found them to be compassionate, altruistic, intelligent, playful and capable of an emotional depth that rivals our own. Therefore, in this book, I refer to them simply as orcas. Who are we to judge them on the basis of their hunting techniques when we, as humans, are capable of so much worse?

Orcas are found all over the world, in every single ocean, some fifty thousand individuals who range from the tropics to the poles, but it is the frigid, fertile waters of the northern and southern hemisphere's high latitudes that they are particularly fond of. Some populations spend their entire lives offshore, while some roam huge stretches of coastline in search of their prey and others prefer to stick to the same location throughout their lives.

The British Isles are blessed with a truly spectacular abundance of marine life. In spring and summer, the seas are rich with plankton blooms that support a huge number of fish, including sharks, and marine mammals such as seals, dolphins and even large whales like minke, humpback and the occasional sperm whale. It is not well known for its populations of orcas, but we are lucky enough to experience plenty of sightings every year.

An increasing number of sightings occur around Orkney and Shetland, in the north and occasionally on the east coasts of Scotland too. These are predominantly of a community known as the Northern Isles Community. Your best bet for seeing orcas is in July or November in Shetland, but there is a chance you could see them in any other month. We also get our 'Viking visitors' or migratory orcas from Iceland, who are predominantly fish eaters but, every summer, some come over to feast on seal pups. About 30 or so individuals are known to split their time between Iceland and Shetland and Orkney, depending on what food they are eating. This represents only a very small fraction of the Icelandic population, and it isn't really known why only a handful make the journey. You are only likely to see them for a couple of months in the summer.

However, there is a small group of orcas present in the British Isles and Ireland who are referred to as our only resident pod and they can be seen throughout the year, predominantly on the rugged and wet west coast of Scotland, in the Hebrides.

Some argue that the Northern Isles Community should now be considered semi-resident or even resident in UK waters, but it is the West Coast Community that has long been considered our only resident pod. Sometimes seen in Ireland, Wales and even England, it is the Hebrides that is their main home.

For those who do not know, the Hebrides is a long archipelago of Atlantic islands off the west coast of Scotland that formed some 525 million years ago in the Precambrian period. This landscape was formed by the ice ages and the great cycles of thaw and freeze. Evidence of its long history is all around, from rocks and boulders which have become polished and smooth through the ebb and flow of tides over millennia to the more recent 4,000-year-old peat bogs that cover much of the landscape. Split into two main groups, the Inner and Outer Hebrides are separated by the Minch in the north and the Sea of the Hebrides in the south. Seventy-nine islands comprise the Inner Hebrides, thirty-five of which are inhabited, including the Isles of Mull, Jura, Skye and Rum. The Outer Hebrides, while separated by only a small sea, seem to be a world away, with their own distinctive language, culture and deeply held religious views. They consist of some fifty substantial islands, with many smaller islands jotted among them. In the inhabited islands of the Outer Hebrides, or *Na h-Eileanan an Iar*, Scottish Gaelic is still widely spoken even after centuries of decline.

Although the Hebrides lie fairly far north at Latitude 57, the same latitude as parts of Sweden, Russia, Alaska and Quebec, they experience a relatively mild climate thanks to the Gulf Stream. Some of the best beaches in the world can be found here, with crystal-clear blue waters, golden sandy beaches and beautiful sand dunes. Just behind the sand dunes, the fringes of the coastline are covered by *machair,* the Gaelic word for the low-lying fertile land that is a characteristic habitat of the Outer Hebrides. If you were to see a picture of the Isle of Harris on a sunny day you could easily mistake it for the Caribbean – an illusion that would quickly end the moment you dipped your toe into the cold waters.

The population of orcas that call the Hebrides home are known as the West Coast Community (WCC). This group has been delighting scientists and whale enthusiasts for years, but we still know remarkably little about them. They spend their days milling around the sparsely populated Inner and Outer Hebridean Islands, in an area of outstanding natural beauty but also where there is some of the harshest weather in the British Isles. These are, at best, only eight members left – four males and four females. Not long ago there were ten, and stories abound of more along with calves back in the 1980s, but no one really knows if the group was once more prolific. Since 2016, only two of the males have been sighted, raising concerns that they may be the only two remaining.

Tracking them down is no easy feat, making their lives something of a mystery. What we do know is largely down to citizen science where members of the public, from diehard orca fans to lucky ferry passengers who chance upon a sighting submit details and photographs to the Hebridean Whale and Dolphin Trust (HWDT), a small but effective charity located in Tobermory on the Isle of Mull. Since their establishment in 1994, they have been collecting sightings, gathering information and running research expeditions to learn about the marine life of the Hebrides – not just orcas, but fifteen other species and counting. They have generated the largest coherent dataset of its kind in the UK, evidence which has been used to establish new marine protected areas, guide fisheries in adopting safer fishing practices, run educational programmes across the Hebrides and help understand the specific challenges marine life faces in this most beautiful place.

They promote mindful engagement with nature and have created the Hebridean Whale Trail, which includes over thirty sites across the west coast and Hebrides where you can watch whales and dolphins from land with little to no impact on marine life. At each site volunteers engage with and educate visitors, and sightings are added to their Whale Track app. It is their work and local connections that have made HWDT the success that it is.

From citizen-science sightings like these, and research expedition data, we can piece together where the West Coast Community is seen, who they are seen with, where they have been travelling to and from, how old its members might be and more besides. It is thanks to the people who help with sightings that we know as much as we do.

The West Coast Community has been hitting the headlines in recent years, grabbing the attention of the world. They have been featured on *The One Show* on the BBC, in countless national and local newspapers and even the US *Smithsonian*. Sadly, the orcas' turn in the media spotlight in 2016 arrived with the heartbreaking discovery of one of their number stranded on a beach after getting entangled in fishing gear. Losing this individual brought the population one step closer to extinction and, with that, their plight became of interest around the world.

My journey to writing this book began in 2014 when I travelled to the Hebrides to join a research expedition on the *Silurian,* the research vessel belonging to the HWDT. Built in 1981 in Seattle, she is a 61ft Skoochum one ketch, meaning she has two masts for the sails, although her motor is used more often than not to keep her speed and direction constant. She has had quite a life, having been impounded for smuggling cocaine from Colombia into Florida.

Redeeming her from this criminal past, she was used by the

BBC in filming the landmark series *The Blue Planet*. The 'first ever comprehensive series on the natural history of the world's oceans', it was narrated by Sir David Attenborough and captivated audiences worldwide with its groundbreaking cinematography and accounts of natural marine events.

Silurian's part involved travelling to the Bahamas to film dolphins, the Azores to film sperm whales and Spain to film pilot whales. In 2002, she headed back to the Azores to get some much-desired footage of bait-ball feeding. The team were running out of time, but at the eleventh hour got lucky and managed to capture some fantastic footage of a feeding frenzy involving shearwaters, yellowfin tuna and Mako sharks. This scene became the climax for the 'Open Ocean' episode.

After a short stint with the International Fund for Animal Welfare in the Caribbean, she was purchased by the HWDT in 2002 and taken to her new home on the west coast of Scotland, where, from April to October, she travels across the Hebrides collecting data on marine mammals, monitoring the use of damaging fishing practices and ensuring that the military testing that takes place twice a year does not impact the whales and dolphins.

When not involved in marine surveys, she is a floating classroom, welcoming on board school children of all ages to learn about marine life by very interactive and immersive means. A permanent crew of three or four dedicate their summers to working on surveys, leaving space for six or seven volunteers to help with data collection. She is the perfect research vessel, having been upgraded in her *Blue Planet* days, and is now equipped with computer systems for logging data and hydrophones (underwater microphones) that are deployed each day, as well as a crow's nest as an additional observation platform.

I climbed to the crow's nest as we made our way to the Isle of Rum to witness a sea so calm it was like a mirror beneath a sky that was deep and cloudless. It was the perfect moment to see the Hebrides from this higher perspective. Having spent the previous eight days and nights with the crew and volunteers, thirty minutes in the crow's nest provided some much-appreciated solitude to reflect, soak in the beauty and connect with nature. I am not one for heights, so going up and coming down were not particularly pleasant experiences, but that view was worth every unwanted heart palpitation.

The adventures started well before I boarded the *Silurian*. After a short flight from London Heathrow, I boarded a train at Glasgow's Queen Street Station and headed to Oban, one of the prettiest of train rides through the mountainous Trossachs National Park.

Travelling through the Western Highlands is an exhilarating and illuminating experience in itself. It was late afternoon when I boarded, so I approached the western shores of Scotland as the sun was setting. The track took us through dense spruce forests and mountains that were shrouded in mist and fog. There was something so calming and peaceful about the journey, a real sense of remoteness that seemed a million miles from the chaos of Heathrow just a few hours before. Between the forest trees came glimpses of the many vast lochs that are scattered throughout the country. Some are lakes, cut off from the sea in the last ice age, others are sea inlets.

It was evening when I arrived in Oban and, after a long day of travelling, I was more than ready to check into my B&B for the night. My sea adventure began the next day when I boarded the CalMac ferry from Oban to Craignure on the Isle of Mull, one of the largest islands of the Inner Hebrides. It was just a short journey of 45 minutes but, of course, I spent the whole time on the top deck scanning the waters for signs of life, practising my skills of observation.

Mull is an important destination for any wildlife enthusiast. The rare white-tailed eagle was reintroduced to Scotland in the 1970s and 1980s, and can often be seen along with golden eagles, otters, whales, dolphins and basking sharks on or around the island.

Its principal town of Tobermory hugs the shore, with buildings painted in bright, bold colours, and is one of the main gateways for exploring the Inner and Outer Hebrides. The harbour is full of fishing boats, but wildlife-watching boats are on the rise as more people become aware of the abundant wildlife of the Hebrides. As I wandered up and down the quirky main road, looking in the quaint shops and tea rooms, it felt very much as if time was running on a different scale. I could feel myself breathing slower, deeper, my body relaxing despite the heavy backpack I was lugging around with everything I would need over the next two weeks. That night, my new teammates and I (nine of us in total) boarded the *Silurian,* had dinner and an introductory lecture on our roles and the wildlife we would likely encounter. Talk of orcas was brief. Clearly no one had high expectations.

The next morning we woke early to make final preparations before setting off with hydrophones deployed, binoculars at the ready and hopes high. Over the course of the day we travelled some 50-odd nautical miles from the Isle of Mull to the Isle of Rum, but had no luck with marine-mammal sightings. Meanwhile, I took any opportunity I could to chat to Tim, our skipper, and learn all I could from him.

Tim wasn't a permanent member of the HWDT team but occasionally skippered for them, and I was glad to have him on our trip. He lived with his family in the Outer Hebrides and knew these waters like the back of his hand. A real outdoor adventurer, he spent his time either mountain biking the trails of the Outer Hebrides, kayaking around the small islands and camping in sheltered bays, or sailing boats like the *Silurian*. Definitely not a man for a regular nine-to-five job in an office. It's fair to say he was (and remains) a bit of an adrenaline junkie, pursuing countless extreme challenges which were fascinating and inspirational to hear about. A bit of a joker who didn't take life too seriously, he was always quick to take charge if the boat ever went off course and had a healthy respect for nature and the power of the ocean. I enjoyed sitting with him at the helm whenever I could, where I told him about my desire to see orcas and that I one day hoped to write a book about them. I found him to have a very calming and relaxing persona, but he was also, clearly, very knowledgeable.

As we passed the Ardnamurchan Lighthouse he told me that, although he hadn't been so lucky, this was a great place to spot orcas and that a high proportion of *Silurian's* citizen-science sightings were made here. There is a visitor centre at the lighthouse, which has the claim to fame of being the most westerly point on the British mainland. From the lighthouse there are commanding views over the inner Hebridean Islands and the Sound of Mull and it attracts many visitors each year between April and October. With so many eyes on the water at this specific spot, it is no wonder that a lot of the sightings come from here. That said, the waters here are deep and fast-flowing, which results in a lot of eddies and upwelling. This creates a nutrient-rich area which attracts lots of fish and therefore a lot of seals and porpoises and they in turn attract the orcas. I didn't take my eyes off the water the whole time Tim spoke, constantly scanning the area for signs of that distinctive dorsal fin.

We were joined by Tom, the first mate, a young and quiet chap who lives on Mull. It was such a joy to talk to them both and hear their stories of life in the Hebrides. They clearly led rich and fulfilled lives and spent most of their summers out on the water. However, it was disheartening to hear that neither had ever seen an orca, much less one of the West Coast Community. If these two, who live here and spend most days on the water hadn't seen one, then chances were that I wouldn't on this short trip. My hopes were quickly diminishing but, ever the optimist, I kept scanning the water looking for signs.

We were blessed with largely good weather on our voyage around the Hebrides, something you do not take for granted in a place where the wind is relentless and the weather can change in an instant. We travelled to the Isle of Rum, the Little Minch, the Shiant Islands, Stornoway, the Isle of Harris, and even to St Kilda some forty kilometres west of the most westerly part of the Outer Hebrides. The scenery was breathtaking, even on the days when the wind blew a gale and the sun remained hidden by dense rain clouds, but it was, of course, when the skies were clear and blue that we enjoyed it most.

Spectacular as the land and seascapes were, we had come here for the wildlife and it did not disappoint. Over 400 sightings of marine mammals in all meant an average of forty each day we were at sea, including countless bobbing heads of harbour and grey seals, harbour porpoises nearly everywhere, and common dolphins, including a superpod of more than a hundred that surrounded the boat and took turns to enjoy a bow ride. What a sight that was: dolphins as far as the eye could see with calves nestled next to their mothers and the sounds of their echolocation blasting out from the hydrophone. We were lucky enough to have an encounter with an unusually friendly minke whale, usually a shy and elusive species. On this particular day though, one decided it wanted to play and joined us for thirty minutes of rolling over to show us its big white belly.

Our strangest sighting though, was of a giant leaping bluefin tuna being chased by a common dolphin off Soay, an islet in the St Kilda archipelago. Quite possibly the last thing we expected to see. Then there were the seabirds that filled the skies around the islands: skuas, shags, kittiwakes, fulmers, eagles and gannets galore. St Kilda alone is home to one of the world's largest colony of gannets, with over sixty thousand breeding pairs. Counting them was a futile task; it was simply impossible to even estimate a number for that many moving creatures. The Hebrides truly is a remarkable place for wildlife.

However, I hadn't travelled here to see kittiwakes and, thankfully, didn't have to wait long to experience what I had really been looking for. On the second day, one of those not so rare grey, bleak days in Scotland, as we passed Skye's Neist Point Lighthouse I saw something in the distance while on watch. A tall, black object appeared from nowhere and disappeared just as quickly. I shouted, 'Sighting, orca!'

Everyone jumped, grabbed their binoculars and started scanning the ocean to see ... nothing. Panic rose in my chest, fear that the orca might have disappeared and that I had lost my opportunity, but mostly panic that I might disappoint a boatload of whale enthusiasts eager to see an orca that was never there.

Panic over ... there it was again, and there was a second. There was no confusing them with anything else, their dorsal fins were so distinctive. As they approached us, I soaked up every second, stealing calm from within when I wanted to scream with excitement, trying to regain some measure of professional composure. I had longed to see these amazing animals up close for years and now the moment had arrived it did not disappoint, though it was not quite how I had imagined.

They were not curious about us at all. They paid us no attention, but I was struck by how big and commanding they were. At around eight metres in length, they were nowhere near as big as humpback whales, which can reach sixteen metres and which I have encountered up close in a kayak. Yet, somehow, they seemed bigger, more Herculean, and there was no chance that I would get in the water with them. Not only is the water freezing cold in Scotland (just trust me on that one), but these two males were powerful and their movements unpredictable. They displayed no signs of aggression, but their power and presence told you not to mess with them. They were simply majestic, elegant and captivating. A bit of healthy respect for a wild animal isn't a bad thing.

For an hour we had the pleasure of watching them as they went about their day. The hour seemed like mere minutes, as it always does when you are fully absorbed in an experience. Then they took a deep breath and swam away.

We didn't see them again that day, nor on the rest of the trip, but that memory will remain with me always. It was my first encounter with wild orcas, but it would not be the last.

Photo identification confirmed that they were from the West Coast Community, a tiny population that is genetically and culturally isolated from other orcas that visit these bountiful and chilly waters. This family does not appear to mix with any others. I say 'family', but it is not really known how they are related. However, given that other populations stay in their family units, it is presumed they have some familial connection. In fact, recent genetic sampling of one of the females, Lulu, has shown that she was heavily inbred, suggesting that they are in fact a family.

It is thought that calves were stillborn back in the 1980s or died in the first few weeks of life. They have not been seen with a calf in the 30 years that they have been formally monitored. Given that they were all adults the first time they were identified by the HWDT, it is now likely that they are too old to breed, too inbred and their bodies too laden with pollutants. This means it is only a matter of time until, one by one, we see the West Coast Community disappear. We are literally watching the end of a bloodline of the oceans' most remarkable animal.

Some might ask whether losing those remaining from the West Coast Community matters. Orcas will still exist in other parts of the world, and indeed in the British Isles, but their loss will be felt along the west coast of Scotland and Ireland. They may be just a few individuals, but they are our marine apex predator and possibly the last remaining individuals of an entire ecotype of orca. They play a key role in maintaining ecological balance and trophic dynamics in our waters and, without them, we could see huge shifts in predator-prey numbers in the coming years.

Losing this family matters to our ocean's health and, chances are if you are reading this book, it will matter to you on a more personal level. Other groups are also dying out, with the Southern Resident orcas (more commonly referred to as the Southern Resident Killer Whales, SRKW) in the Pacific Northwest having received the greatest coverage and interest. Across the globe, populations of orcas are struggling to cope with the marine environment as it is today. For 11 million years they have thrived but now are languishing. For many of these populations, little is known and even less reported.

Our time to learn from the West Coast Community is ending and, once they are gone, we will forever be haunted by unanswered questions such as: *Were they really a family? How did they get to be so isolated? Why couldn't they breed? Why did we never see them all together? What role did humans play in their life? What role did we play in their demise?* Questions like these might apply to other small populations of orca, and indeed other species, and finding the answers might enable us to reverse their decline and demise. Sadly, likely not in time for the West Coast Community.

In telling their story we inevitably consider many aspects in the lives of orcas generally – their social structure and life history, communication, impacts of pollution and noise, and what we can all do to help them. As I dug deeper though, I discovered that this sorrowful, majestic ten have their own tale to tell with a powerful moral behind it, one we should not choose to ignore. Their stories range beyond the Hebrides and through time, and touch on much that is relevant to our human lives and the dangers we share with all the world's creatures in a time of climate crisis and increasing pollution. The demise of our ocean's apex predator is a stark warning of what may befall other species and ecosystems if we do not learn how to coexist in harmony. My hope is that through these stories, my research and interviews with the people who know them best and orca experts, you might be touched and motivated to do something for our aquatic friends.

This book comes in two parts, in chapters that do not follow catalogue order, which might jar slightly with my more academic colleagues. Instead, artistic licence has been taken, confusing the order in the hope of clarifying the wider story. The second part describes that wonderful, revelatory voyage on the *Silurian* and, I hope, gives the feel of the Hebrides when they are first met – the taste or *blas*, as it is expressed in Gaelic. The first part, that leads into it, tells the story of John Coe, Nicola, Comet, Moon, Lulu, Floppy Fin, Puffin, Aquarius, Occasus and Moneypenny. It feels right to share it while we still have the West Coast Community, our orcas of the Hebrides.

PART ONE

The West Coast Community

JOHN COE

Residents, transients and everything in between

WE START OUR STORY WITH JOHN COE, the first member of the West
Coast Community. He was added to the HWDT catalogue in
1992 with the name John Coe and the numeral code 001. However,
he was named John Coe much earlier, in the 1980s, when the Sea
Watch Foundation (SWF) team, led by its Director Dr Peter
Evans, encountered him. SWF is a national marine charity
focussed on improving conservation for whales and dolphins
around Britain and Ireland. The team had been on the *Marguerite
Explorer*, a charismatic traditional 120ft sailboat, following a group
of orcas around the Hebrides.

The owner and skipper of the boat, Christopher Swann, or
Swanny, saw this huge mature bull and thought he should have a
nickname. Swanny was particularly keen on nicknames, teammates no exception, with Dr Evans being dubbed 'Shocker' by
him. Dr Evans didn't divulge the reason behind his nickname and
I thought better than to ask.

Swanny had been reading the book *Mile Zero* by Thomas
Sanchez, which is about a freed slave, John Coe, who becomes a
student of the sea. The nickname stuck and is still how the most
famous member of the group is known. Dr Evans said that 'it was
a fitting name for this great wanderer of the ocean who must
know its waters better than most'.

So, for many years before the HWDT was founded, John Coe
and his community were known to researchers as well as locals of

the Isle of Mull. Daniel Brooks, a former Sea Life Surveys employee, said he grew up on the island in the 1980s being told about a group of thirteen orcas that were regularly seen, and that there were reports of calves in the group.

We can be confident that the John Coe we know now is the same animal as back then, as he is easily the most distinctive not only of this group but also of other orcas seen in Scotland.

Should you be lucky enough to encounter an orca in the wild, there are several features you can use to determine who you are looking at. The easiest identifier is their dorsal fin, the large triangular fin on their back, which is likely to be the first thing you see as it will be jutting out of the water and, at nearly two metres in height for a full-grown male, it is pretty hard to miss. Much as our fingerprints are unique to each of us, orca dorsal fins are unique, with different shapes, sizes, slopes, nicks, scars etc. John Coe's dorsal fin is easily recognisable, with its curved top leading to a vertical drop on the rear side and a large, very noticeable notch near the base.

There are many other identifiers we can use to tell who we are looking at: the eye patch, being the white area behind the eye, and the saddle patch, being the white area at the base of the dorsal fin, can tell us a lot about what population the individual might belong to, as we shall see later in this chapter.

In 2015, the HWDT spotted that John Coe had acquired a new and very obvious telltale: a large bite out of his tail fluke. With the very obvious notch in his dorsal fin and this unique bite out of his tail, it would be near impossible to confuse him with another whale.

After a close inspection of a photo and consultations with marine experts from various organisations, it was decided by the HWDT that it was most likely a shark bite. Over twenty species of shark are known to visit Scottish waters, most fairly small and all harmless to humans. Some are present all year round, such as the lesser and greater spotted catshark, spurdog, common smooth hound, and others seasonally such as the basking shark, tope, porbeagle. Others are seen only on rare occasions such as the sofa shark, angel shark and the bluntnose six-gill shark. John Coe's bite, though, is pretty sizeable and must have come from a similarly sizeable shark, which rules out most of those.

The basking shark (*Cetorhinus maximus)* is the second largest of the shark species, growing to a maximum length of 12.2 metres, but most only grow to between six and eight. I say 'only' but that is still bigger than a great white. The basking shark is a gentle giant that moves through the sea with its mouth agape, feeding on plankton in the water column and is particularly abundant in the Hebrides in summer months when plankton blooms occur. As filter feeders they are toothless. Therefore, it is safe to say that the attack on John Coe was not from the basking shark.

In 2021, John Coe was spotted off the coast of Cornwall, a very unusual sighting for the West Coast Community, so it does seem his roaming takes him into southern waters. Maybe he came across a great white shark on one of his adventures to warmer climes and got in a tussle, but which species of shark took a mouthful of John Coe must be shelved as a case of whodunit.

Biting does not all go one way though, as some orcas have an appetite for sharks, such as the offshore ecotype of British Columbia or a population in New Zealand that hunts rays, a type of shark in the elasmobranch family. Shark skin is really quite special. Unlike other fish that have scales, sharks have dermal denticles, which are like hard, grooved teeth, with a central pulp cavity, dentine and an outer layer of enamel. If you run your hand lengthwise down their bodies you will find that the scales are smooth. Go the other way and you will find that they are incredibly tough and will snag your skin. Orcas that are known to eat sharks often have teeth so worn down that they are left with only gums and are unable to feed any longer.

The orca that live in British waters are not known to eat sharks although, since we know so little about them, it must be regarded as a possibility. So, how did John Coe get this bite out of his tail? Was he trying to eat a shark when he realised that he had bitten off more than he could chew? Did the shark bite back? Or was there a particularly bolshy shark that thought it would have a go? If so, I can't imagine the shark coming out of that altercation in a good state.

Altercations with sharks have also been documented for Bigg's killer whales that prey on gray whales, particularly gray whale calves. Every spring, Pacific gray whales (*Eschrichtius robustus*) migrate up the coast from Baja California past British Columbia up to Alaska and the Arctic seas.

This is a journey of 12,000 miles (round trip) taking several months, and is one of the longest migrations of any mammal on the planet. The calves are born in the warm, calm waters over the winter months in Baja, California. While safe from predators, it is lacking in food. The mother will have fasted since leaving feeding grounds in the Arctic many months ago and now must eat, and lots, if she is to nurse her calf sufficiently.

Both the mother and baby must make the long journey to the bountiful food supplies in the north Pacific, swimming as quickly as they can to evade predators and get the mother to food but limited in their speed by the calf, which has to work hard to keep up with Mum.

Keeping close to the shoreline, mothers protect their babies in the kelp and in shallow waters from the orcas that patrol the coastline at this time of year.

Mothers will fiercely protect their precious cargo and will fight back, but even with their best efforts, it is thought that up to a third of all gray whale calves migrating up the Pacific coastline fall victim to orcas each year.

However, a gray whale, even a calf, is too much to eat in one go, even for a family of orca. The carcass sinks to the bottom, but given the kill usually happens in relatively shallow waters of about 20 metres deep, the carcass is only a short dive away.

Researchers have observed orcas returning to the same place the next day to resume feeding on the carcass, using these shallow waters as a kind of fridge, if you like, for their food. The pungent smell and the oil slick from the decaying body acts as an 'X', marking the spot for several days, allowing the orcas to make repeated trips.

However, the carcass quickly attracts attention from other hungry predators. Sleeper sharks, a deep-water shark that is rarely ever seen in shallow waters, make their move, entering the larder to feast on this fortuitously easy meal. It appears that the smell of the feast is just too much to resist and so under the cover of darkness, they come up from the depths of the cold Alaskan sea and join the feeding party.

While it has not been documented, it wouldn't take too much of a stretch of imagination to envisage the scene where the orcas are interrupted by a group of sleeper sharks while tucking into their hard-earned dinner. The orcas worked tirelessly for hours on end to take down the beast, presuming it would feed their family for the coming days. But the sleeper shark in a feeding frenzy is on a mission, and a few orcas are not going to stop it.

A fight ensues and bites are given, on both sides. Perhaps a surreptitious bite to the tail fluke, as far from the jaws of the orca as possible.

While this kind of 'food storing' behaviour has only been documented for a newly recorded population of transient killer whales in Alaska, there is a possibility that similar habits are seen elsewhere in the world, including in the Hebrides.

The West Coast Community are known to occasionally prey on minke whales, another rorqual whale, although much smaller than a gray whale, but still likely to be too much to eat in one go.

Perhaps our West Coast Community feed on these larger whales and store them in shallow waters for future feeding. The carcasses in turn attract sharks and lead to a feeding frenzy of the top predators in Scottish waters. Maybe this is how John Coe got the bite out of his tail fluke. A bit of a leap perhaps, and the real reason remains a mystery.

John Coe is known to get around: not as a ladies' man, but geographically. The Hebrides is his main stomping ground, but he has been sighted off the coast of Ireland as well as Pembrokeshire and Anglesey in Wales. While its members are mostly seen in the Hebrides, it appears that the West Coast Community travels the west coast of both Scotland and Ireland. There have only been a couple of sightings on the east coast of Scotland.

When John Coe was observed off Peterhead in Aberdeenshire in 2013 it was the first time in twenty years he had been spotted in the east. On 5 May 2021, he created quite the media frenzy when he appeared off Land's End in Cornwall with his last remaining pal, Aquarius. There are several Facebook groups dedicated to orcas in Scotland and some solely for the West Coast Community, not to mention the HWDT social pages, with thousands of people following reports of orca movements around the UK and Ireland. Those living near the coast might head out to sea or to good vantage points on land in the hope of getting a glimpse. Those living further afield eagerly await updates posted online. Just before the sighting in Cornwall, tensions were rising in the community of watchers who follow their movements as John Coe and Aquarius hadn't been seen in over seven months. Everyone was preparing themselves for the possibility that they had passed on.

In populations such as the Southern Resident orcas of the Pacific Northwest, if an individual isn't seen for one year then it will usually be declared 'deceased' by the Centre for Whale Research. This not-for-profit organisation have been studying this population for 45 years, and can declare this with more certainty than us due to their extensive knowledge of the orca population. If the rest of the family is seen several times that year, but an individual is missing each time, something is amiss. Photographs from drones are taken most years, helping to identify which orcas may be looking in poor condition or showing signs of starvation. Then if an individual that was already looking emaciated isn't seen the next year, it adds to the body of evidence that an individual has died, or is 'pushing up the kelp' as it is sometimes put.

However, with the West Coast Community, it is quite common for individuals not to be seen every year and, without drone footage, we cannot get an impression of the state of their health. Therefore, the HWDT do not make such declarations and can only assume after several years of no sightings that they *may* have passed on and are presumed dead. With the risk of their extinction now so close, any year without a sighting may well signify their passing.

With this sighting in Cornwall there came a huge collective sigh of relief. It was also the very first time that any members of the West Coast Community had been seen in Cornish waters. In fact, it is the only time that any of them have been seen in England. What an absolute treat.

To everyone's surprise, just nine days later, on 14 May, the same pair were seen again in their home waters off the Isle of Skye. In nine days, they had covered some 895 kilometres, an incredible rate of travel at about a hundred kilometres a day, but the story doesn't end there.

On 13 June 2021, the dynamic duo were seen again, this time off the coast of Dover in the English Channel. This came as quite the surprise to the open-water swimmers who were swimming the English Channel for charity and, I have to admit, I too would have been shocked to find myself suddenly and unexpectedly accompanied by a pair of orcas.

Then it gets even more curious … It is thought (but not confirmed) that the pair were seen seven days later near John O'Groats in north-east Scotland.

Just what was going on? For decades, this community remained largely loyal to the Hebrides and, less so, to the west coast of Ireland. Now, as their numbers dwindle, they seem to be roaming far and wide. Did they do a complete circuit of the British Isles? Are these two males searching for others of their kind? Are they looking to find a new home? Or have they always roamed this far, but organised observation and social media have made sightings and reporting so much more frequent? We are faced with another mystery about this group but, it is safe to say, they keep our attention and the watcher community is holding its collective breath for the next sighting, always hopeful that there will be another plot twist.

John Coe is just one of at least fifty thousand orcas that live in the world's oceans, having adapted to almost every environment and a wide range of food sources. While currently recognised as a single species *(Orcinus orca),* there are many scientists who believe they should be classed as several different species or at the very least as a species complex, meaning that, while they look and appear to be so similar they are, in fact, not the same but are closely related groups which are, at present, difficult to define. That is because around the world, and indeed in the same geographic areas, there are populations that do not feed in the same way, do not breed together or interact in any way. They likely don't even understand one another linguistically, using different dialects. Morphologically, ecologically and culturally distinct, they nonetheless look similar to the untrained eye.

It is currently accepted that there are ten ecotypes, that is, ten ecologically distinct 'races' of orca. However, these ten by no means cover all of the known populations. Five of these ecotypes live in the southern hemisphere and five in the northern hemisphere. To understand how John Coe and the other members of the West Coast Community fit into this global organisation, let's take a quick look at the various ecotypes.

Southern hemisphere ecotypes

The pristine waters of the Antarctic have, for millennia, remained largely unchanged. The winds blow fiercely and relentlessly and for half of the year the region is plunged into darkness. The other half brings constant daylight. The water temperature can range between −2°C and 10°C, incredibly cold but still much warmer than the land temperature of about −50°C which the penguin populations endure. While Antarctica has been a place of stable environmental conditions, that is now rapidly changing and it is warming at a rate faster than the rest of the planet, resulting in less ice and shifting ecosystem patterns.

In the austral summer, as the ice packs thaw and the water temperatures begin to warm, the Southern Ocean begins to teem with life. Upwelling of nutrient-rich water from the depths starts the phytoplankton bloom, which is then followed by a huge abundance of krill, a tiny crustacean that baleen whales consume at an impressive rate.

A humpback whale can consume about five thousand pounds of krill each day. The krill and fish population are so concentrated in these waters that marine mammals flock here to feast on them. Today, around sixty thousand humpback whales travel from northern Australia where they have spent their winter mating and raising their newborns. Food is scarcer near the tropics and so they head south to the bountiful waters off Antarctica to gorge and replenish their fat stores.

It sounds like a lot, sixty thousand humpback whales, and thankfully it is, but it is a far cry from the numbers that were seen before they were hunted to near extinction. Commercial whaling in Antarctica started in 1904, much later than in other regions such as Scotland, which started way back in 1753. Whalers must have thought they had hit the marine-mammal jackpot. With long summer days as the sun barely dips below the horizon, the whalers have plenty of time to get their haul. Mysticeti, otherwise known as baleen whales (since they have no teeth but rather baleen plates with hairs to filter krill and small fish from the water) were the target. The blue whale was the most lucrative whale to catch due to its immense size, being the largest animal to have ever lived. When the blue whales had all but disappeared, the whalers moved onto other baleen whales such as the fin whales, sei whales, minke whales and humpback whales.

In the years between 1904 and 1986, when commercial whaling was banned by the International Whaling Commission (IWC), whalers all but decimated the whale population. Blue whales used to number up to 300,000 individuals pre-whaling but dropped to just three thousand individuals. Humpback whales had historic populations of close to 100,000 but declined to just 200 individuals. Since commercial fisheries were banned, populations have been recovering and it is a joy to know the abundance of these majestic creatures returning to these waters.

While commercial whaling has been banned under the IWC, large whales continued to be removed from the Southern Ocean by the Japanese for their highly controversial 'scientific research' purposes. Sadly, in 2019 Japan left the IWC and has again resumed commercial whaling. We can only hope that the populations have recovered enough to withstand this increased pressure.

Orcas, strictly a member of the dolphin family and not a whale, with teeth and not baleen, were not really a target species in Antarctica, or anywhere else where large-scale commercial whaling was taking place. They are much smaller than the baleen whales the whalers were after; they are quick, agile and very intelligent. However, as in other whaling or fishing hotspots, they would likely have been a nuisance and killed if they got in the way of a hunt. Today, some 25,000 to 27,000 orcas roam the freezing cold waters of Antarctica, accounting for half of the world's populations. There are five ecotypes in the southern hemisphere, each with their own unique characteristics.

1. Antarctic Type A orca

This type looks like the orcas we are accustomed to seeing in movies and on television, with a large black and white body measuring up to 9.5 metres in length. Generally seen in relatively small groups of twenty individuals or fewer (not that twenty orcas seen together seem like a small group), each year, in the austral summer (i.e., when it is winter in the Hebrides) they migrate back to the rich feeding grounds of the ice-free waters of Antarctica to hunt minke whales (*Balaenoptera bonaerensis*) and elephant seals *(Mirounga leonina)*. They have even been seen to chase penguins, but never actually to catch one.

Unlike the other populations of orca found in the Antarctic, the Type As are only seen in the open ocean as they tend to avoid pack ice. When the temperature starts to plummet and darkness descends, the minke whales leave the Antarctic and the Type As follow. During the austral winter, it is thought that they migrate to lower latitudes, to New Zealand and perhaps even the tropics. Good evidence for this is found in bites from cookie cutter sharks, so called because they feed by gouging round chunks of flesh from their prey. These sharks are only found in warmer tropical waters. Clearly, the Type As are not big fans of freezing cold, long, dark winters. Them and me both.

2. Pack ice orca (Large Type B)

A large, two-toned grey and white form with a dark cape pattern and a very large eye patch. That's right, not the black and white we are used to. The cape, which is also something that we are not accustomed to seeing in the movies or television, is bounded by a distinct line that runs from the eye patch along the back until it meets the saddle patch below the dorsal fin. Above this line, the skin is usually darker than below, making the whales look as if they are wearing a dark cape as they glide through the water.

Individuals often have a yellowish appearance, thought to be from diatoms on their skin, a tiny plant that is part of the plankton soup. As their name implies, these orcas are usually found in the pack ice, where they hunt Weddell seals (*Leptonychotes weddellii*) and the occasional humpback whale calf. You might have seen a video of Sir David Attenborough describing these guys hunt. The poor seal is seen resting on the ice floe, only to be spotted by a group of orcas who carry out an impressive, co-ordinated attack. By swimming in unison towards the seal and diving at the last second they create a wave that rocks the ice floe until, after several attempts, the wave washes the seal off the ice and into the water, making him an easy pick for the clever orcas.

As the viewer you can't help but root for the seal, but an orca family has got to eat too. That said, researchers have observed them feeding on Weddell seals after what can only be described as careful butchering. Once their prey had been meticulously dismembered, they ate their favoured parts and abandoned the rest, clearly not that desperate for food. This is apparently wasteful, but leaves food for scavenger species to survive on.

These orcas hit the headlines after observations suggested that they might, just might, have met their match in humpback whales who have been seen preventing, for some reason, the orcas getting to their prey. With the humpbacks' apparent permission, exhausted seals, desperate to evade the orcas, have been seen to rest and take shelter on their large bodies. While the orcas might go for a whale calf, taking on a giant full-grown humpback is another matter. Definitely possible, but a lot more effort.

Why the humpbacks protect other species in this way is not known, particularly as there appears to be no benefit to them. Possibly it is a wonderful display of altruistic behaviour rarely seen in nature. Then again, if we are to ascribe human characteristies to cetaceans (whales, dolphins and porpoises), perhaps it is a personal vendetta against the ocean's most successful hunter after the loss of a calf or sick relation.

3. Gerlache orca (small Type B)

Like the pack ice orca, the Gerlache type also has a dark cape, a large white eye patch and a yellowish tint from diatoms. It is a bit smaller, but a big difference is that, while its preferred prey is unknown, it has been seen to feed on penguins, both the gentoo (*Pygoscelis papus*) and the chinstrap (*Pygoscelis antarctica*), with a particular penchant for the breast muscles.

The breast muscle on birds is the largest muscle, accounting for up to 20 per cent of total body mass so, if you are going to target one muscle, breast is best. As their name suggests, they can mostly be found in the Gerlache Strait, which is the channel separating the Palmer archipelago from the Antarctic peninsula. If, like me, your knowledge of the intricate geography of the Antarctic is a little rusty, grab a map and take a look. The coastline, and indeed the whole territory, is a source of constant fascination.

4. Ross Sea orca (Type C)

The smallest of all the known orcas, adult males only reach about six metres in length. Their grey and white body has a dark cape like the others in Antarctica, which is again and for the same reason covered in a yellowish tint. However, this one is distinguishable by its very narrow and slanted eye patch, which is a bit like the eye patch on the West Coast Community.

These orcas spend most of their time deep in the Ross Sea pack ice hunting for fish, particularly the Antarctic toothfish (*Dissostichus mawsoni*). The toothfish, unfortunately, is heavily targeted by the fishing industry, whose predation in turn has an impact on the orca as less food is available. The toothfish is a popular prey item for other species such as seals, penguins and other whales, so there is a lot of competition for this one fish.

The Ross Sea orca numbers appear to have declined in recent years, which could be because they have moved away in search of new feeding grounds or, possibly, because their numbers are decreasing with the food supply. Sadly though, the most likely cause is the second of these, with overfishing of their main prey species driving their decline.

5. Subantarctic orca (Type D)

The remaining type of orca found in the Antarctic has only recently been described. However, genetic analysis shows that it has been separated from the other types for nearly 400,000 years. This is to say that, although inhabiting the same waters, these orcas have not interbred with their Antarctic neighbours in all that time. They have only been spotted a dozen times, but are easily identified by their tiny eye patch, rounded head and a dorsal fin that is more pointed than the others and swept back rather than standing upright.

They tend to stay in the open ocean, which is why it is so much more difficult to study them, and it isn't known what they feed on. Fish seems most likely, as they have been reported as stealing from longlines. Understandably, they are not popular with fishermen.

They were originally identified from a group in 1955 that washed up on a beach in Paraparaumu on the North Island of New Zealand and which scientists realised were very different from other orcas. It wasn't until early in 2019 that they could be confirmed as existing. Dr Bob Pitman, a marine ecologist at the American National Oceanic and Atmospheric Administration (NOAA) and Marine Mammal Institute at Oregon State University, is a leading authority on Antarctic orcas who managed to get together the funds and a small team of five. Heading into the Antarctic, their aim was to locate these elusive whales from only the evidence of a handful of stories of encounters and some photos.

They were in luck and had a short but golden three-hour opportunity to see and film this ecotype, confirming not only their presence but also their uniqueness. Unlike the other orcas in the region, this ecotype does not appear to like the most southerly colder waters of Antarctica. They also appeared to be very friendly, approaching the boat and even killing a seal right before them. Dr Pitman said they are the largest undescribed animals left on the planet. What a thrill it must have been for that team to see them first-hand.

Northern hemisphere ecotypes

There are five ecotypes in the northern hemisphere, three in the Pacific Northwest and two in the North Atlantic. Most of what we know about orcas has come from research conducted in the Pacific Northwest. It is a place of big nature running from the US states of northern California, Oregon and Washington and then up into British Columbia in Canada. The climate is coastal temperate. With the Rocky Mountains on one side and the mighty Pacific on the other, the region gets a lot of rain and remains relatively mild. But up into Alaska the Arctic takes hold, and snow and ice become a central part of the climate.

Ancient coastal temperate rainforests comprised of redwood cedars, Douglas firs, hemlocks and spruce trees grow majestically all along the rocky coastline, reaching unimaginable heights. In the morning, the forest is enshrouded by mist and fog, the warm breath the trees have produced in the night as it meets the cold night air. It is such a beautiful illustration of the living nature of our planet. You can literally see the earth breathing. Walk through the forest and you feel its pulse as you bounce over the soft padded floor covered in pines and dark, earthy moss. So much life is hidden here. It is a place of quiet solitude and yet if you stop to listen you will hear the sounds of life all around: babbling brooks and streams that criss-cross the landscape, tiny insects scurrying through the detritus on the forest floor, a cacophony of birds in the trees, not to mention the elk, moose, bears and wolves that make their home here.

These forests extend right out to the water's edge, the trees growing on the rocky outcrops that dominate the coastline in this region. The waters are teeming with marine life, with productive kelp forests to rival the forests above the water's edge. It is a place where the forest and ocean act as one ecosystem, each dependent on the other for survival, a place where you feel simultaneously insignificant and central to the natural cycles. Salmon is the key to this environment. They feed the seas, the land and the sky, being the main source of food for orcas, bears, wolves and eagles. These creatures are not just ecologically important but culturally very significant to the people of the First Nations.

This region has been home to First Nation communities for thousands of years, living in harmony and balance with nature. But in more recent times, since European settlements and explorations, we have been exploiting its abundance of natural resources by damming rivers, felling forests and using aggressive fishery methods which have altered the ecosystems drastically.

In contrast to the Pacific Northwest, in the North Atlantic regions where orcas dominate, there are no ancient temperate rainforests. The Northern Isles and the Hebrides are characteristically treeless and barren. It is still an area of outstanding natural beauty with wild weather, with plenty of rain and driving winds from across the Atlantic, but the mountains and coastline are bare in comparison.

Orcas are seen mostly in Scotland and up into Iceland and the fjords of Norway and even up to the Barents Sea. While the climate around the Hebrides is relatively mild thanks to the Gulf Stream, it is certainly not so mild when you move north into the Arctic circle. It is a dream of mine to visit the population of orcas in the spectacular fjords of Norway in the winter when the steep walls of the long narrow fjords are covered in snow. The whiteness reflects the light show of the Aurora Borealis, the Northern Lights, that play in the sky while large pods of orcas gather in the freezing waters to feed on herring. The North Atlantic, primarily the west coast of Scotland, is of course home to our West Coast Community, which we do not believe have ever ventured as far north as Iceland or Norway. If you are incredibly lucky, you might just get to see both orca and Aurora Borealis at the same time in the Hebrides.

1. Resident orca

The resident orcas of the Pacific Northwest are easily the most studied of all orcas, having been the subject of continuous research since the 1970s. They have predictable distributions, particularly in the summer when their movements are dictated by the movements of their favoured prey, the Chinook salmon. Split into two populations, the Northern Residents spend their summers in Johnstone Strait off North Vancouver Island, and the Southern Residents spend the summer and autumn in the Puget Sound and Salish Seas on the border between Canada and the United States.

Resident orcas live in extremely close family units, called matrilines, with sons and daughters remaining with their mother throughout life, sometimes with three or four generations living together every single day of their shared lives. They are often seen in groups of twenty or more, and in the summer, until only recently, they were seen almost daily. However, sightings in the summer months are now less frequent as they roam greater distances in search of food, but more on that later.

I was so lucky to see and study individuals from both populations in the summer of 2016. In fact, my passion for orcas started after reading the book *Orca: The Whale Called Killer*, written by Erich Hoyt in 1981 about the Northern Residents.

Hoyt eloquently described his summers in the 1970s, spent following a resident pod of orcas in the beautiful, rugged and wild Johnstone Strait. His close encounters with the ocean's apex predators led to his gradual realisation that they were not the fearsome, voracious predators many believed them to be. Instead, they were intelligent and gentle, even compassionate. He told many stories of calves being born and old ones lost, intimate moments shared and the pressures the community faced.

I was hooked and just had to see them for myself. So, in 2012 I travelled to Telegraph Cove, a quaint old boardwalk village consisting of only a handful of houses built on stilts over the water. From there I kayaked with my boyfriend (now my husband) down Johnstone Strait to just outside the Robson Bight Ecological Reserve, which is now a sanctuary for the orca in part thanks to Hoyt. For four days, we paddled up and down these waters, constantly scanning the horizon for a glimpse.

At one point we thought we saw dorsal fins in the far distance. My heart began to race, both through excitement that I might finally get to see them up close and through an element of fear, for that very same reason. Have no doubt, these are large creatures, robust and strong. They could snap our kayak in half without a second thought, flip us up and leave us helpless and at their mercy if they wanted.

But deep down I knew this wouldn't happen. There has never been a single incident in the wild of an orca causing intentional harm to a human. They are frequently near boats and kayakers and seem much less interested in us than we are in them. We are just another thing to navigate and, at best, momentarily inspect.

As it turns out, the distant dorsal fins were nothing more exciting than some floating logs, something that is common in this heavily forested (and deforested) part of the world. My disappointment was palpable; however, I think there was a large sigh of relief from Adam behind me! Given he was not a massive fan of being on the water, I was lucky he had agreed to the trip at all. I am not so sure we would have had the same reaction to seeing orcas from our little plastic kayak.

My disappointment was somewhat assuaged by the many encounters with humpback whales we had in those four days. Especially memorable was a humpback mother and her calf having early morning speech lessons. With the mist hugging the forest-lined rocky coast, the sun was just starting to break through the dense clouds and the sea was mirror-flat, not a ripple.

We stood there in the cold early morning among the drift-wood on the shore, blankets over our shoulders, warm cups of coffee between our hands, as our eyes and minds slowly woke up from the deep sleep that you only seem to get when immersed in nature. None of us uttered a word.

The only noise came from the strange, dinosaur-like sounds across the water of a mother calling and then the noise repeated by her calf. At first, the calf's noise was nothing like the mother's, but with practice and perseverance it gradually got closer and closer to the same pattern, if at a higher pitch.

While we didn't see any orcas, my curiosity had been piqued, and on a return trip in 2016 I was lucky enough to see representatives of both Northern and Southern Resident orca communities. It was every bit as magical as I had hoped.

2. Bigg's orca (transient)

These orcas used to be called transient orcas as they roamed freely in the waters of the Pacific Northwest, staying in one area for only a short time. Unlike the residents, they do not have a predictable pattern. Their other, more widely accepted, name is Bigg's orca. So named in memory of the late Dr Michael Bigg, the founding father of orca research, who dedicated his life to their protection.

Bigg's eat other marine mammals such as Steller sea lions *(Eumetopias jubatus),* harbour seals *(Phoca vitulina)* and harbour porpoises *(Phocoena phocoena),* but they will also hunt sea otters *(Enhydra lutris),* minke whales *(Balaenoptera acutorostrata)* and calves of large whales.

All of these animals have exceptional hearing and are quick enough to evade an approaching predator. Having to adapt to this fact, Bigg's orcas have evolved into stealth hunters who hang out in groups of three or four individuals. They swim silently in these small groups to stay inconspicuous and sneak up on their prey. Large groups would be too noisy and since they share their kill between the group, sticking to a small group not only makes a kill easier but also ensures more food for each orca.

Because they are not seen every year, their life history and family units are not as well understood as the residents with whom they have not interbred for thousands of years. In fact, they do not interact at all, tending to avoid each other, even to the extent of chasing the other group away. Mostly though, each is indifferent to the other.

In my research for this book, I travelled once again in 2016 to British Columbia, but this time to Tofino on the western side of Vancouver Island. Here, the coastline is even more wild and rugged than Johnstone Strait, buffeted by strong Pacific winds that grow in strength as they sweep across the ocean from Japan.

I came to Tofino to work with the Strawberry Isle Marine Research Society and learn about the Bigg's orcas they have been monitoring since the 1980s. Their diet of other marine mammals defines so much in the lives of Bigg's orcas, and their distribution, communication and social structure are all likely determined by this factor. The West Coast Community in Scotland are also marine-mammal eaters and so my instinct was that they are likely to be much more similar to the transient orcas we see in the Pacific Northwest than the residents. Learning about the Bigg's orcas of Vancouver Island paved the way for me to a better understanding of our own ill-fated ten.

3. Offshores

The third type of orca in the Pacific Northwest is the Offshores. Of these we know very little as they spend their time far out to sea, where they are difficult to observe. What we do know is that they are smaller than the first two types, growing to a maximum of about 6.7 metres, and live in groups of between one and two hundred individuals. In these large groups they travel great distances, ranging from Alaska to southern California.

Like the residents, they are acoustically active and spend a great deal of time chatting away to each other, in direct contrast to the Bigg's, who are silent for most of the time so as not to alert their prey. The Offshores appear to feed mostly on shark species, whose rough skin wears down their teeth until only stumps on the gum line are left.

An offshore orca with the ID 0120 washed up near Tofino in 1997. After months of painstaking and smelly work, her rotting flesh was removed and her bones bleached and preserved. Her skeleton was reassembled and now travels across Canada as part of an educational programme run by the Strawberry Isle Marine Research Society. Many children will never have seen such a large animal, and the opportunity to play with her bones and learn about her in such a tactile way is an invaluable teaching tool, providing a learning experience they are sure to not forget.

The last two ecotypes are found in the Atlantic and are known as

Type 1 Eastern North Atlantic and Type 2 Eastern North Atlantic. By far the better understood is the Type 1, of which there are several discrete populations. It's complicated, I know, but bear with me.

4. *Type 1 Eastern North Atlantic*

Type 1s are fairly small, growing only to about 6.6 metres in length. Mostly found in Iceland and Norway, they feed on Atlantic herring *(Clupea harengus)* and to a lesser extent, Atlantic mackerel *(Scomber scombrus)*. Hunting in impressive, co-ordinated movements, they herd the herring into a tight ball and drive them to the surface before using their tails to smack the ball hard, stunning the fish and leaving them as easy picks when they fall away, a hunting technique known as carousel feeding. Birds will join in the frenzy from above, and other marine mammals, such as humpback whales, will join from below, taking huge mouthfuls of tiny silver fish.

The Type 1s' teeth are often found to be worn down, and it is thought that, because they suck up the fish, over time the pressure from sucking takes its toll on their dentition. A scientist called Dr Andy Foote, arguably the leading authority on British orcas (from the University of Aberdeen) along with colleagues from the University of Copenhagen, have performed tests using stable isotope analysis to determine exactly what orcas are eating and it appears that, for some, fish constitute only part of their diet. Stable isotope analysis is based on 'you are what you eat'. Most elements exist in two forms or isotopes, each with a different mass, a light one and a heavy one. The relative abundance of each type of isotope varies among food in a food web and the ratios are incorporated in the tissues of the animal higher up in the food chain. Therefore, stable isotope analysis can identify what the orcas, in this case, have been eating.

While other fish-eating orcas appear to feed exclusively on fish, some of the Type 1s from Iceland spend their winters eating herring in Iceland then come to Shetland and the north coast of mainland Scotland in the summer to feast on seal pups. At least thirty individuals are site-faithful, such as Mousa and her family, Hulk and Nótt, and will return in the spring and summer to feed predominantly on seals along the coastline. This prey switching is quite unusual for orcas.

However, they are not the only orcas to prey on seals in Shetland. Most sightings from Shetland and Orkney are of a different Type 1 community called the Northern Isles Community, such as the pods 27s, 64s and 65s, who appear to be rather generalist hunters feeding on other small marine mammals such as seals and harbour porpoises. They can be seen in any month, but July and November especially, and there is a good chance you can spot them just by standing on the shoreline as many come close to land to hunt. Sightings of the Northern Isles Community are becoming more frequent and many now consider the members to be resident to the Northern Isles. The increase in sightings and young being born into this population is very encouraging and exciting.

For a long time it was also assumed that some of the Scottish sightings of orcas were of the visiting Norwegian population. However, there is scant evidence of this and only one sighting has ever been matched between the two populations. In 2018, off Vatersay in the Outer Hebrides, a group of unknown orcas were spotted and photographed by the HWDT on the *Silurian*. Who they were was a mystery but they were added to the Scottish Killer Whale Catalogue. It wasn't until 2021 that they were seen again, this time in Norway where they were unknown to the local scientists. Using the Scottish catalogue, a match was made, providing the first and only evidence of orcas moving between the two locations.

Genetic sampling has shown that the Type 1s share the same maternal lineage; this is to say that they have a common ancestor from way back, and are not the mixed population they were once assumed to be. Rather, they are distinct communities within the Type 1s. A recent study has shown that the call types, that is the vocal calls they emit to communicate with other pod members, differ significantly between the Norwegian population and those from Shetland and Iceland. Additionally, photo-identification supports the idea that there is now no mixing between these populations.

5. *Type 2 Eastern North Atlantic*

Dr Andy Foote noticed that not all orcas in the eastern North Atlantic were the same. Using the extracted DNA and teeth from museum specimens he established that there is another type living in the eastern North Atlantic, the Type 2. Only recently formally recognised, the designation is based solely on one dead specimen, an orca that was stranded in Sandwood Bay in north-west Scotland in 2007, and from old museum samples.

Much larger than the Type 1s, they can reach nine metres in length. They are also easily distinguished by their white eye patch which slants backwards and downwards, unlike the Type 1s, or for that matter any other ecotypes except one in the Antarctic. They also have sharp teeth, unlike the Type 1s after suction feeding small fish.

This type does not eat fish, but hunts other cetaceans such as whales and dolphins, especially the harbour porpoise (*Phocoena phocoena*). Stomach-content analysis has showed that minke whales (*Balaenoptera acutorostrata*) can also be on the menu. On the surface of it, it seems that the West Coast Community are Type 2.

Not a lot else is known.

That was a lot of information so here is a quick recap. There are ten 'ecotypes' or ecologically distinct populations of orcas so far identified. Five live in the southern hemisphere, centred around Antarctica, and five in the northern hemisphere. Three are in the Pacific Northwest, which is where most of our knowledge of them comes, and two in the north-east Atlantic.

These ten ecotypes do not encompass all known orca populations, so our knowledge of ecotypes and population organisation is very much a work in progress. It does appear though, that we are witnessing something called speciation. Orcas have been on our planet for a VERY long time and populations have, for the last few hundred thousand years, become isolated from each other; they are socially, culturally, and genetically different. Their lives remain ever changing and evolving, and that is pretty cool!

Whether resident, transient, Type 1 or Type 2 Eastern North Atlantic, or none of these ecotypes, John Coe, and indeed the West Coast Community, is often described as resident to the Hebrides and so could be confused with the fish-eating 'Residents' of the Pacific, but they are much more like the Bigg's orca in that they eat other marine mammals and travel quite long distances.

John Coe typifies what it is to be 'transient'. Rather than hanging around a good feeding patch, he moves from place to place in his search for food. Other populations of orca around the world will stay for months at a time in a good feeding spot, bringing their family and usually returning to this same feeding spot year on year as long as fish are plentiful.

For the resident population in Norway, it is herring (*Clupea harengus*), for residents in British Columbia it is the Chinook salmon (*Oncorhynchus tshawytscha*) and in Antarctica, the Ross Sea killer whales hunt Antarctic toothfish (*Dissostichus mawsoni*). The return of these fish to a particular location is often predictable and the orcas know exactly when and where to be. Their return to the Pacific Northwest in the summer months is celebrated by First Nation communities as it signals the start of the salmon spawning season, meaning not just plentiful food for the orcas or bears but also for humans.

However, the food that John Coe and his ilk are after is not fish. It is other marine mammals. This group is known to eat a variety of other warm-blooded animals, mainly harbour porpoises, and known to take larger prey such as the minke whales which frequent the Hebridean Sea. Minke whales are gentle giants with a shy, nervous disposition who rarely approach boats or display extravagant behaviour such as the breaching and tail slapping of their more gregarious relative, the humpback whale. Minke whales are the smallest of the baleen whales, but they are still a large whale, reaching a length of nine metres.

Often, when orcas are taking on such large prey, they harass and chase in turn, swapping roles when the lead chaser gets tired. This way they conserve their energy while forcing their prey into exhaustion. Our orcas in the Hebrides are mammal eaters and they are smart. John Coe and his family must use stealth and surprise to catch their prey off guard because, once their prey is alerted to their presence, they scarper. Seals will haul out, porpoises become even more elusive and minke whales will hide among kelp and keep close to rocky outcrops . . . and John Coe will move from place to place in search of inexperienced, naïve prey.

We know the West Coast Community orca are bigger than the Type 1s from Shetland, Iceland and Norway, that they have an eye patch that slants backwards and sharp teeth. So, it is unlikely they are another population within the Type 1 ecotype. On the surface, it seems that they are the Type 2 Eastern North Atlantic population, and many now refer to John Coe as a Type 2, but with no genetic evidence, it is not so clear.

To add another layer of confusion, scientists are now questioning whether the classification of Type 1 and Type 2 ecotypes for the North Atlantic orca accurately represents the current understanding. Our knowledge of orcas in our waters is ever evolving, much like the orcas themselves.

While there appear to be many similarities with the Bigg's ecotype, genetic analysis has shown that the West Coast Community is more like the orcas of the Antarctic than those in the Atlantic or Pacific.

In fact, it is postulated that the West Coast Community is really a splinter group from an Antarctic Type A population that long ago tried to carve out a niche for itself in a different area. If so, and sadly, the strategy does not appear to be working well for them. Maybe they got lost and just couldn't find their way back south again. What we are seeing could be the last of the group and ancestral line, desperately hanging on in their final years.

NICOLA

Matriarchs and menopause

WHILE JOHN COE IS THE BIG MALE that many believe to be the leader of the group, the true boss of this family is (or was) actually likely to be Nicola (003), the third orca to be catalogued, who was first named in 1992 by the HWDT. She was most frequently sighted with John Coe and Floppy Fin, but she hasn't been seen for many years.

Photo-identification studies have shown that the West Coast Community tend to form strong bonds in pairs, a male and a female, called dyadic associations. Nicola tends to favour the company of Floppy Fin, Moneypenny favours John Coe, Occasus favours Aquarius, and Puffin favours Comet. Lulu and Moon didn't seem to form such strong bonds.

Some believe these associations are of mother and offspring, meaning Nicola could be Floppy Fin's mum, and some scientists believe that she could also be the mother of the other males, including John Coe. In orca communities, females are the head of the family, the matriarchs, and responsible for leading. However, the species is rather exceptional in that it is one of only a few where sons and daughters will stay with their mothers for life. The only other marine species with this characteristic is thought to be the long-finned pilot whale (*Globicephala melas*).

The scientific term for this social structure, where animals remain with the group they were born into, is called 'natal philopatry'. Among the resident orcas of the Pacific Northwest, both sons and daughters stay with their mothers for life in this matrilineal structure. Each matriline can then form up to three or four generations of maternal descendants of a single living female and her children, grandchildren and even great-grandchildren. Members of a matriline will stay in near constant acoustic contact with each other for life, rarely spending any time apart.

A pod of resident orcas is made up of several matrilines and can be as many as twenty members strong, all linked to the principal matriarch in some way. She will lead her family around the waters that she is intimately familiar with, to all the best feeding spots, teaching them where to go and what to be aware of. She will also teach them all the vocal dialects of their ancestry, with each matriline having calls that are unique to that family.

However, they have another set of calls that they share with other groups of the same clan. In this way, the family members can easily identify to which wider pod, or acoustic clan, they belong. This is particularly handy when trying to keep your family in order but, more importantly, it lets all the whales know who is not from the family, and therefore with whom they can mate without risk of inbreeding.

The fancy term for this strategy is 'natal group exogamy' and it is a smart solution which allows long-lasting close family bonds without risk of mating with your mother or sister. Scientists try hard not to anthropomorphise, but some have found it hard not to relate this social structure to that of human tribal societies.

Orca specialist Dr Lance Barrett-Lennard, in his PhD thesis, compares orcas to small tribal societies who live together in close-knit communities but who maintain population segregation to avoid the downfalls of inbreeding. I had the pleasure of supporting Lance and his team at the Marine Mammal Research Centre in Vancouver in 2016, where I learnt so much from him. He has some amazing stories to tell of his field trips in Johnstone Strait.

It appears from a very recent study that our West Coast Community is extremely inbred and so natal group exogamy doesn't appear to be a strategy that works for them. This is no doubt due to a lack of suitable mating partners. They are just too culturally and socially isolated and so are left with no choice but to mate within their group. It is postulated that it is their history of inbreeding that has left them unable to breed.

While resident orcas show these incredibly strong family bonds, it is not the case for the transients of British Columbia. Bigg's orca, a type of mammal-eating orca, have a much more fluid relationship with family members and are generally not referred to as pods. In fact, the Strawberry Isle Marine Research Society likes to refer to them as 'gangs'.

While sons and daughters both initially stay with their mother, some males will decide to roam the waters on their own and simply leave. Occasionally these lone rangers might pair up with another male for a while before heading off again on their own. Daughters tend to stay with their mothers until they have their own offspring. While residents have big families of twenty or more travelling together, this many orcas in one place are hardly inconspicuous and, since they hunt other marine mammals with exceptional hearing, they must try to be discreet.

It has been shown that the optimal group size for hunting prey such as seals and harbour porpoises is three. It doesn't take any more to hunt down such small prey and, also, with additional numbers the energy efficiency per prey item diminishes. In reality, the groups are usually a bit larger, such as four individuals, as there will often be a calf or immature orca who will not be of great assistance and is still learning the ropes. Witnessing the hunt is of high importance for these animals as so much will depend on their learning the tricks and ways of their family.

Given that there aren't any calves in Nicola's west coast family, the group sizes for a hunt are smaller than is desirable, but it takes only two to track down a harbour porpoise.

Back in 2006, Daniel Brooks was a guide for Sea Life Surveys on the Isle of Mull, out on the water with a boatload of enthusiastic whale watchers when one shouted that they had seen an orca. When Daniel and Richard Fairbairn looked, they initially thought it was a humpback whale as the dorsal fin was small, not the upright colossal two-metre dorsal fin they were expecting, until it came up again. It wasn't a humpback whale. It was Floppy Fin (we will get to him later in the book). Another came up, and this time it was Nicola, matriarch of the family.

Incredibly, it looked like a baby was situated between the two. Excitement levels were high. Two of the West Coast Community with a calf. Had Nicola had a baby after all? But, as they surfaced again, the 'calf' was seen to be in Floppy Fin's mouth and it was no calf, it was a porpoise. When they disappeared underwater to share their meal, excitement quickly turned to horror then to disappointment and finally appreciation.

The boat was only a hundred metres away, pretty close, but it wasn't the gory sight you might imagine. Bits of blubber floated to the surface which attracted lots of seagulls, but Daniel and Richard managed to retrieve a piece. It still had muscle and skin attached, so a positive ID of porpoise could be made.

Then a rather strange thing happened. Curiosity took over, along with the opportunity to connect with the whales on another level. The people on the boat divided and ate pieces of the blubber. They were not overly appetising, but it was an incredible experience to share a meal with the West Coast Community. The hunt was not spectacular. There was no thrashing, throwing of the prey, or aerial displays, just a quick kill and a meal shared among family, and friends.

*   *   *

We know that in larger units of Bigg's orcas, once a daughter has her own offspring, the chances are that she will split from her natal group, and this new group will grow in size with each offspring until they, in turn, break away.

However, even when daughters have left their mother, they usually stay in close contact and meet regularly. Daughter and offspring will often frequent the key feeding areas that she was introduced to by her mother, and, in this way, a strong cultural inheritance is passed down the generations.

The West Coast Community was rarely seen together as a group of ten, which would be typical of other resident pods. Instead, they had an average group size of just three. Given this average, their unpredictable movements and fluid social organisation, it is likely that they are more comparable to the Bigg's orca, which is to say transient.

Regardless of being transient or resident in nature, it is still likely that Nicola, as the possible matriarch, is the most important individual in the family, responsible for guiding the group and keeping them together.

There are some exceptional matriarch stories that I simply must share.

Granny (J2), was a Southern Resident orca in the Pacific Northwest and possibly the most famous of all the matriarchs. It is possible that she was born in 1911, three years before the First World War. She was the matriarch of the whole of J clan which consisted of some seventy-eight animals in December 2016 and was reduced to just seventy-four by September 2020. What this means is that all the orcas frequenting the San Juan Islands around Washington State and off Southern Vancouver Island are related to her in some way.

At an estimated 105 years of age, in June 2016, she was seen still leading the group to find food. That summer I had been working as a visiting scientist with the Marine Mammal Research team at Vancouver Aquarium, writing blog posts and analysing drone footage of the Northern and Southern Resident orcas. On one of my days off, I joined a whale-watching boat departing from Granville Island and heading south towards the San Juan Islands, where reports had come in that the Southern Residents were nearby. Not only did we get to see the Southern Residents but among them we got to see Granny. She was right there at the front, leading the whole of J pod and some of K pod as well, thirty-three animals in all, in Haro Strait off San Juan Island. Sadly, just a few months later in December 2016, she was reported missing presumed dead.

It is always a spectacle to see the Southern Residents, who seem to be the most flamboyant of the Pacific populations, delighting in full body breaching, tail slapping and general tomfoolery. They also took a good look at us by spyhopping whenever they could, positioning themselves vertically in the water with their head and, more importantly, their eyes out of the water. This allows them to take a look at what is going on in our world above the water line. The Southern Residents certainly create good photo opportunities. Aside from being the subject of many a tourist photo, Granny was also a film star as she was one of the whales used in the filming of the movie *Free Willy*.

A really impressive female, she remained in almost constant contact with all her family and extended family for over a century, imparting the knowledge she must have held on the best feeding spots, where the best Chinook salmon were and how to navigate the complex maze of islands and coastline.

Let's take a moment to consider what she must have witnessed in her lifetime.

Mass-scale whaling came to British Columbia in 1905 and commercial whaling continued on a global scale until 1986 when the IWC banned it. Huge numbers of marine mammals were harpooned and dragged to shore to be cut up for their bones and blubber. Orcas never became a whalers' favourite. That was the right whale, so called because whalers deemed it the 'right whale' to catch as it is slow, easy to harpoon and delivered a nice profit.

The oceans in those days would have been filled with a cacophony of calls, whistles and clicks from a range of giants. With numbers reduced to near extinction, it must have become eerily quiet. Although orcas were not targeted for whaling, they were generally considered a pest by whalers and fishermen alike and were often gunned down to prevent them from interfering with operations, or to stop them taking fish from the line. Sometimes they were shot just for fun.

In the 1970s, when the tables began to turn, orcas went from being one of the most hated animals on the planet to one of the most loved. They were targeted for the captivity industry, and nowhere was the capture effort as intense as off the coast of British Columbia and Washington State. Granny's family was particularly 'favoured'.

In the summer months, Resident orcas follow highly predictable patterns close to land where huge numbers of salmon aggregate, which made it easy to predict where they would be and launch an attack. During the captures in the 1970s, it is thought that up to fifty Southern Residents were captured although, as you might expect, the industry didn't keep logs of the whales taken or count those killed in the chase.

Years later, several orca hunters are now orca activists brave enough to share their experiences. Being the social, family-oriented animals they are, when a calf was targeted, family members rarely left it to fend for itself, often trying to free the calf or make things difficult for the captors. Harpoons, guns, explosives, you name it, were used to separate the family and isolate the baby.

The adults, too heavy to transport, were just a nuisance and, once killed to make life easier or accidently in the chase, the captors needed to get rid of the evidence. The captured calf would then witness its family members cut open, filled with rocks and sunk to the seafloor. The horrors of these events were witnessed only by those on the boats and the surviving orcas. Given that population numbers before this industry started were not known, we will never truly know how many were killed, but we can be sure that those who actually reached marine parks represent just a fraction of those lost.

The captive industry was particularly interested in the young as they were small and naïve, making them easier and cheaper to move from the sea to their awaiting concrete tanks, wherever in the world that might be. Sometimes to aquariums in Canada and the US, others as far afield as Windsor Safari Park, five miles from where I grew up! The site is now home to the hugely popular Legoland.

Essentially, a whole generation was taken from Granny's family and, nearly forty years on, the impact is still evident in a lack of breeding whales.

Being over fifty years old then, Granny wouldn't have been attractive to the trade. It would have taken a lot of effort to move a whale that big and heavy who probably didn't have long left anyway. Knowing what we do now about the level of empathy these creatures have, their emotional intelligence and their incredibly strong family ties, far exceeding humans in some cases, it must have been beyond heartbreaking to see so many of her family members taken, and many more die in attempted captures.

Given their amazing memories, I doubt that she ever forgot, and yet she has never approached humankind with any hostility. Few humans would show the same grace. Since 1975, when the last capture of wild orcas took place in British Columbia, Granny's family has struggled to recover. Now Critically Endangered, there were only 75 as of July 2023. Although many more than the ten of the West Coast Community, it is a far cry from what it was.

The Southern Residents now have to navigate the home waters they share with an ever-growing number of boats which are not only a physical navigational challenge, but also noisy. I mean *really* noisy. Orcas have eyesight as good as ours, both in and out of water, but it is their truly impressive acoustic capabilities that they rely on most. A talkative bunch, the Southern Residents are in near constant communication with each other. What they are talking about is, obviously, not really known to us, but it is usually assumed to be along the lines of 'fish over here', 'I am here, where are you?', 'found a great kelp patch to play in' and 'watch out for that boat!'

I am convinced that this assumption is a massive over-simplification, and that they are really communicating much more than we will ever understand.

They use echolocation to create a picture of the underwater world, using a string of sonar pulses directed out of the melon on their forehead. When a beam strikes an object it bounces back to be received through a special fluid in the lower jawbone which directs it to the inner ear where the vibrations are processed through nerve impulses to create an image, in much the same way as we hear. Additional noise in the water from boats therefore impacts their ability to communicate and find food.

One study measured their communication when cargo ships pass by, sometimes miles away. Sound travels far in the water, and sometimes the orcas would just talk loudly when they went past. Other times they simply fell silent, like we do when an express train passes through a station without stop-ping. Too loud to continue their conversation, they would just cease.

The long-term implications of this are unknown, but with more and more cargo ships and busier waterways, we may be sentencing these vocally active, sentient creatures to a life of silence.

Another notable matriarch is the late Stubbs (A1), a Northern Resident in the Pacific Northwest off Vancouver Island.

Stubbs is thought to have been born in 1927 and she died in 1974 at the age of forty-seven. While the length of her life was not particularly notable, she is very special in that she was the first orca to be individually identified and given a systematic name, hence A1. She was a favourite among pioneering researchers such as the late Dr Michael Bigg, along with Dr John Ford, Graeme Ellis and Erich Hoyt.

From Hoyt's stories of Stubbs in his book (*Orca: The Whale Called Killer*), it is easy to understand their affection. She was instantly recognisable with a deformed dorsal fin, likely the result of an encounter with a boat propeller which cut off the end to leave a jagged trailing edge. If a clear favourite of early researchers, it was not for her beauty. She was generally considered to be fairly ugly, with a big rotund belly, a strange neck and that famously deformed dorsal fin. She was certainly not the most attractive of orcas. In fact, she was given the nickname 'Stumbellina, the ugly duckling'.

She was friendly though, often approaching boats and show-ing an interest in humans. She had just two known offspring, a daughter Sophia (A36) and a son Hardy (A20, also called Sturdy). Hoyt wrote that with each returning summer's field-work he would desperately scan the waters, looking out for Stubbs.

One year, it seemed that she was struggling to keep up. She was always at the back of the group, bringing up the rear rather than leading the way, but her family would wait for her and let her rest before setting off again in search of salmon. Often, she would be seen chatting away with Nicola, another grandmother within the family (not Nicola of the West Coast Community), I imagine in much the same way my grandmother used to while away the hours gossiping to other old ladies at the local coffee shop. Eventually, in 1974, when Hoyt and his colleagues returned to Johnstone Strait, she simply never showed. Her family were there, but not Stubbs, and it could only be concluded that she too was now 'pushing up the kelp'.

I have to admit that when I read that story I cried, even knowing that her family went on giving us fascinating new insights into their lives.

The pod was subsequently renamed the A36s, as Sophia was now the matriarch and leader. Sophia only had one daughter who sadly died at the tender age of two, but she also had three sons. It was presumed that when she went her sons would separate, as there would no longer be a matriarch to hold them together. However, the boys stayed together and have never been seen apart. The fact that the strong family ties continued even after the passing of Stubbs and Sophia is testament to the values and traditions that Stubbs instilled.

The A1 pod is still one of the healthiest in British Columbia. They have few incidences of members dying young and have had many healthy calves. They are also the most approachable whales in Johnstone Strait with apparently no fear of boats or humans and will come right up close out of curiosity before carrying on with their day.

I have also been fortunate enough to see Stubbs's descendants and experience the joy of watching them sleep as a group, breathing in synchrony until, one by one, they wakened from their nap. Stubbs was the start of more than forty years of research. She inspired a generation of researchers to dedicate their lives to understanding, and therefore protecting, these creatures. We owe her a lot.

Granny and Stubbs have been studied extensively over the last forty years. We know all of their children, grandchildren, great-grandchildren, even the ones that didn't make it. Scientists can now even tell which females are pregnant, and are gaining more understanding about rates of miscarriage. The distinctive calls that these matriarchs use in their families is also known and, listened to almost daily by scientists and citizens through the summer months, teach us so much about these fascinating animals.

Although we know very little about Nicola, our matriarch from the West Coast Community, by using a few assumptions and our knowledge of orcas in other parts of the world, like Granny and Stubbs, we can piece together her life with a bit more confidence.

We know that females tend to become sexually mature at around the age of fifteen. Once pregnant, they have a long gestation period, between fifteen and eighteen months, nearly double that of a human pregnancy. They give birth to calves about 2.6 metres long who weigh up to 160kg. Usually, females will give birth to between three and five calves in their lifetime, with a gap of three to ten years between, before hitting the menopause and taking on the role of doting grandmother. Male orcas on the other hand generally don't become sexually mature until much later, usually when they are twenty to twenty-five years old. They reach adolescence before this, at around twelve to fifteen, when their dorsal fin sprouts into the huge tower we associate with the big bulls.

If Nicola is the mother of John Coe and Floppy Fin as some believe, then it is likely that she is at least fifty-six years old. Given that John Coe was known by researchers such as Dr Peter Evans in the 1980s and was fully grown then, complete with a large sprouted dorsal fin, his estimated birth year is around 1965, if not before. Nicola would have to have been born in 1950 if she was to be fully mature by 1965. Females have an average lifespan in the wild of about fifty years. However, as we know from Granny, they can live well past that, even past a hundred. While Nicola would now be postmenopausal, she may still have another thirty years of life ahead, based on the assumption that she is the mother of John Coe.

Humans are one of the few species on the planet to experience the menopause, something that most women don't particularly look forward to. Menopause in other species is very rare.

Usually, as a female mammal ages, her fertility declines and she dies. Even the African elephant, which lives a very long time, and as a grandmother takes on the different role of looking after the young ones, can still reproduce. To live for a long time after the ability to reproduce has passed is unusual and only three other animal species are known to experience the menopause: orca, the short-finned pilot whales (*Globicephala macrorhynchus*) and the false killer whale (*Pseudorca crassidens*). However, evidence is mounting that several other species may also exhibit post-reproductive lifespans.

Resident female orcas have been shown to go through the menopause at thirty to forty years, but will continue to live for many more, even decades. I wonder if they are aware of the physi-ological shift that will alter their role in society. Perhaps they also experience hot flushes or the emotional rollercoaster that female humans go through.

There must be some benefits to the inclusive fitness of the group to have post-reproductive females. As we see with Granny, knowledge is a large part of this. Over the years they acquire a vast wealth of knowledge of their environment that they pass to future generations. Orca grandmothers are often seen babysitting their grandchildren and helping their daughters with their parenting skills. It is possible that the reason some captive orcas are less than doting mothers is that they simply did not have older generations to show them how.

Furthermore, research has shown that post-reproductive females are actually critically important to the survival of the group. In the year following a mother's death, sons older than thirty years are over eight times more likely to die than if their mother was still alive. If they are less than thirty years, they are only three times more likely. Daughters seem to fare a little better and are perhaps more self-reliant than the males, but may still experience three times the normal death rate.

Given that males can sire offspring much later than females, it is important for the group to look after the males to ensure the family genes are passed on. Males also live much shorter lives than females and so generally have less time to learn, making them more dependent on the matriarch to lead them. Male orcas really are mummy's boys.

It is thought that post-reproductive females act as repositories of ecological knowledge and, in times of short food supply, will lead the group to better feeding grounds. This is especially important in El Niño years, a natural climatic phenomenon that occurs in two-to-seven-year periods, resulting in unusually warm surface ocean temperatures in the eastern equatorial Pacific. This is the 'warm phase' of the El Niño Southern Oscillation (ENSO). These warmer waters interfere with nature both on land and sea and affect the diversity and abundance of marine life by reducing fish stocks, including the salmon the resident orcas feed on. When food is plentiful, the matriarch might let others take turns in leading while she takes a step back. This way the next generation gets to practise its skills while she helps out with the grandkids. However, in years of limited food supply, the matriarch will always lead.

Over 750 hours of video footage have shown that this is the case for the Southern Residents in years when the Chinook salmon supply is low. They also show that males tend to stick closer to the mum than the daughters. Orcas are very generous, particularly the grandmothers and mothers, and will share most of their meals with their family, but the mothers appear to give special treatment to their sons by supplying them with extra food, perhaps because they are larger and have higher energy needs. As the females mature and welcome their own offspring, they become less dependent on their own mother, but the males tend to need life-long care. It is not just her immediate offspring that benefit though.

In what is known as the 'grandmother effect', where children raised in the presence of their grandmother are more likely to survive, in the case of orcas they are more than four times more likely to survive. When a grandmother ceases to breed, she is able to focus more on leading the group, sharing out the food, and helping her own children and grandchildren to hunt and pass on their cultural ways. The grandmother benefits them all. By helping her grandchildren to survive, she enhances their own chance to reproduce, thus passing on the family genes. This strategy is rarely seen in nature.

Grandmothers pass on other important life lessons by engaging in sexual encounters with younger males, perhaps as a form of sex education, preparing them for when they are mature and go off to continue the family's genes. Male-on-male encounters are also not rare. 'Sword-fighting', yes, it is what you imagine, has been seen on many occasions as these playful creatures frolic in the water. Is it a form of homosexuality, or are their relationships just more fluid than ours? Regardless of whether this is homosexual behaviour or just play, we know they are incredibly tactile, and touch forms a huge part of their life.

In the summer of 2016, I spent several months working with researchers at the Vancouver Aquarium, identifying individual whales from both the Southern and Northern Resident populations in photographs taken from drones. I spent almost three hundred hours looking through thousands of aerial photos and was always amazed to see how frequently they rubbed against each other, swimming so close that their big, black bodies touched, and how often they nuzzled each other. Not just the mothers and the calves – all family members engaged. Touch and the need to constantly reaffirm their close bonds seems to be as important to their existence as the food they hunt.

The reports of menopause relate to resident orcas in close kin-structured societies who remain together for life. In transient populations where the offspring eventually leave their mother to start their own families, her constant guidance and support are not available. Therefore, it may not be the best option to terminate reproduction. We just do not know enough yet about menopause in this ecotype, or in the West Coast Community, to be sure they go through the menopause and cease breeding. Given that the group is rarely (if ever) seen as a whole, it is unlikely that Nicola is key in leading it to better feeding spots.

There is a possibility that Nicola, despite being in her seventies, may still be able to reproduce, which gives us hope that she and other females in the group may still be able to have offspring.

However, there hasn't been a confirmed sighting of her for many years, and, sadly, we may have to accept that she is no longer with us. John Coe and Aquarius head out far and wide these days. Perhaps they are a little lost without their matriarch.

COMET

The trailblazer

COMET (005) WAS first identified by the HWDT in 1998, he has been seen by them many times since, including by me. He was one of the two orcas that I saw back in 2014 when I was on the *Silurian,* making him one of the very first orca that I saw in the wild. While on that day he was with Aquarius, before then he was often seen with John Coe and they appear to be rather adventurous, having frequently been sighted near Ireland. His tall dorsal fin has a distinctive wave and a small nick towards the tip, which has won him a lot of media attention in recent years. Comet is the second name he has been given. Others may know him as Dopey Dick.

On the cold, grey day that was 8 November 1977, an orca made the headlines when it appeared in the River Foyle in the city of Derry, in Northern Ireland. He was rather unfairly nicknamed Dopey Dick, a sly, literary dig for being a bit like Moby Dick but not so smart for swimming into a river. However, Moby Dick was a sperm whale and therefore different to an orca.

The *Journal* newspaper ran the headline '*Huge whales in the Foyle*' on 8 November and his presence was a frequent feature in the news over the next week. The Foyle Fisheries Commission confirmed that he was a whale and about twenty feet (six metres) long, still not identifying him as an orca.

Given how distinctive they are with their black and white

markings, robust bodies and huge dorsal fins, orcas are one of the easiest marine mammals to identify. Still, it was probably the first time most people in Derry had seen one. It was also baffling to scientists what he (correctly presumed to be a male) was doing in the river. Theories included chasing salmon and a fault in his 'radar system' due to parasites.

We now know that the so-called Dopey Dick feeds on other marine mammals such as porpoises, dolphins and minke whales, and that salmon is not usually on the menu. In all the research conducted on similar populations in British Columbia (the transient or Bigg's orca), larger mammal-eating orcas have never once been seen to show any interest in fish. Transients, captured for SeaWorld and other aquariums, appeared to be at a loss about how to even eat a small slippery fish, preferring to starve. Training is often successful, but there have been times, particularly in the early years, when orca simply died of starvation. So, it is unlikely that he was in the river chasing salmon.

After two days in the River Foyle, he showed no signs of wanting to leave. Since being far from home and fresh water was probably unhealthy for him, there were attempts at rescue and the army was enlisted. Using a boat, a whistle and half a side of beef, the soldiers tried to lure him out of the river and back to sea, where he belonged. He was not interested. When coaxed as far as the bridge, he decided to swim in the opposite of the desired direction. It seemed he had other plans and wasn't going to be easily distracted, especially by a side of beef.

Why they thought beef would be appetising to an orca is a bit baffling, especially if they thought he was chasing salmon. A cow really isn't their usual food choice, although there is a hearsay story of one eating a deer.

By this time, huge crowds had gathered by the River Foyle, trying to get a glimpse of this rare and incredible being. I spoke to one lady who was just a child at the time, but even now remembers her mum taking her and her younger brother and sister in their school lunch break. Nor was this her only visit, as she was taken with her class to stand on the bridge in the hope of just a glimpse. The thrill of seeing the orca has remained with Gillian across all these years.

Teachers were keen to make the most of this unique outdoor educational tool, and perhaps as eager themselves. Whole classes were taken out of school to walk along the river.

Rumours began circulating about where he came from, what he was doing and the mischief he was causing. Rumours even, of him overturning one of the army boats that were trying to lure him out to sea.

If this did happen, it would most certainly have hit the headlines, and it would not have been the first such incident. There have been cases where lonely orcas have been, shall we say, playful with boats. Luna, a young male off Vancouver Island, spent several years growing fond of his floating visitors, often nudging boats, spyhopping to get a close look inside the noisy floating boxes and playing with the jets of water coming off the propellers. It was playing with boat propellers that brought him to his untimely, but anticipated, death.

While Dopey Dick brought much that is good to Derry, in 1977 orcas were not yet seen by the public as the intelligent, gentle creatures we now know them to be and the excitement he stirred at the dreary tail end of the year could not last.

On Saturday 12 November, four days after his arrival, he left. He simply turned around and left the river, never to be seen again, and what befell him has remained a mystery. This story was a real highlight in the lives of the people of Derry and it has been told and re-told in the decades since. In fact, he is somewhat of a Derry legend. There is now even a brewery carrying his unfortunate and unfair nickname: 'Dopey Dick Brewery Co' with a cartoon whale as its logo and the slogan: 'We brew our own, it's a whale of an ale.'

The story of Dopey Dick and his Irish adventures is still told and was a highlight at the Return of Colmcille Festival in 2013, which celebrated both Saint Colmcille, also known as Saint Columba, and Derry as the first UK City of Culture. The poster for the festival states, 'Meet Dopey Dick, he came, he saw, he liked it so much … he couldn't leave'. Then again in October 2016, a wooden replica was made of him by the Helen Hamlyn Centre for Design, Royal College of Art. During the festival, community members were invited to write onto pieces of fabric that would form the 'skin of the whale', messages of hope and wishes for the town, clearly associating him with good times.

His story has survived as, in a time of hardship, he gifted the city a sense of hope and much-needed joy, and I think it is a beautiful thing that a brief and unexpected visit from a marine mammal can leave such a lasting legacy and with such profound and beneficial impacts. His memory still evokes these feelings in a city that has since experienced much regeneration.

This wasn't the only time this orca made headlines though. After almost forty years, in 2016, an old black-and-white photo of him was uploaded onto a Facebook group, likely related to a Halloween event planned on the Foyle with the arts college. Scientists from HWDT and Dr Andy Foote, the leading authority on the West Coast Community, recognised the distinctive dorsal fin and matched it with recent photos of Comet.

It turns out that Dopey Dick didn't go far, but returned to his home waters of the Hebrides after his short stay in Ireland. He has since returned, but never caused quite the commotion as in 1977 when he swam up the Foyle. Ironically, Saint Columba was an Irish Saint who set up a monastery on the Isle of Iona in Scotland where Comet has often been sighted. So it is pretty apt that the town of Derry chose this festival, The Return of Colmcille, to re-tell the story of when Comet came to visit.

Comet is not the first marine mammal to enter a river. In fact, there are species of dolphin who live their lives exclusively in river environments, such as the Amazon river dolphin, the Indus river dolphin and the Ganges river dolphin. The Yangtze River Dolphin is now, sadly, extinct. All of these species are under increasing threat as rivers become more polluted and degraded. However, they are very different from their marine relatives and there are no species of whale that live in rivers. In fact, it is rather unusual for large marine mammals to spend any time at all in a river.

In 1985, a twelve metre-long humpback whale entered San Francisco Bay, where he was given the name Humphrey. He stuck around for a while, but after four days headed into the fresh water of Sacramento River. He swam sixty-nine miles from the ocean until, after several weeks, his skin started to grey and he became listless. He was dying.

The community worked hard to get him back to sea, but the usual herding techniques did not work. They tried playing orca sounds on underwater speakers to scare him out of the river, but that did not work. They followed this with a Japanese fishing technique called 'Oikomi' where a flotilla of boats makes unpleasant noises on the water surface. Again, it failed.

A humpback whale researcher, Louis Herman, remembered that humpback whales have incredibly complex and beautiful 'songs' of repeated and predictable sounds that carry vast distances in the ocean. He suspected that the best bet might be to play them but, for this to work, they would need serious manpower to lure Humphrey all the way back to the sea.

Thankfully, with the help of an expert acoustician, Dr Bernie

Krauss, equipment loaned from the US Navy and a private yacht donated for the procedure, Humphrey followed the sounds of his fellows out of the river and back to sea. As soon as the water became more saline, he started vocalising and showing signs of excitement, much to the relief and celebration of the many spectators who had come to witness this event. On 4 November 1985, almost a month after his entry into the Sacramento River, Humphrey swam under the iconic Golden Gate Bridge and back out to the Pacific, but ...

... in 1990, he returned!

This time, he got himself beached on a mudflat on San Francisco Bay and once again required the assistance of Dr Krauss and Louis Herman to get him back out to the Pacific where he belonged. Thankfully, this was the last time he is known to get himself into trouble. The last sighting of Humphrey was in 1991 in the Farallon Islands, near San Francisco.

San Francisco Bay has another, more recent, humpback whale story. A mother and calf entered the Sacramento River in early May 2007. The mother, named Delta, had bad lacerations on her back estimated at two to three inches long and six inches deep. Luckily it was not deep enough to damage the muscle layer and she was not bleeding. Experts thought it unlikely that she was at any risk from these injuries. However, Dawn, her calf, also had a laceration, this time on the lateral side. This cut was deep, had penetrated the muscle tissue and was bleeding.

Lacerations like this can occur when whales stray too close to the propellers of boats, sometimes leaving minor scratches but sometimes multiple deep injuries which can prove fatal. Had the mother moved the calf into the relative safety of the river to recover? They would have been migrating from their tropical breeding grounds to their feeding grounds further north, near Alaska.

Orcas are known to patrol these waters and to work as a group to separate a calf from its mother. The kill then is an easy one, particularly when the calf is injured and vulnerable. Being in the river would have protected her from potential predators, but fresh water for an extended period is unlikely to be good for marine mammals. Government officials and staff from The Marine Mammal Centre (TMMC) in Sausalito, just outside of San Francisco, felt it best to guide the pair back to the ocean.

Having rescued over 23,000 marine mammals, TMMC were well placed to provide assistance to Delta and Dawn. Concerned for their health and the risk of infection from their wounds, the team administered antibiotics, the first time antibiotics had been given to free-swimming wild whales. But tactics that were successful with Humphrey, such as playing humpback whale songs, were not successful with this pair. Delta and Dawn decided for themselves when they were ready to leave and were last seen swimming out from San Francisco Bay on 29 May 2007, almost a month after they entered.

Had the cuts healed sufficiently and were they able to make it to their feeding grounds? That remains a mystery. It is not known if they entered the river because of the wounds, to evade predators or because they were lost. We will never know but, hopefully, there was a happy ending for this mother and her baby.

In April 2018, excitement levels rose on the west coast of Scotland. A pod of orcas had been seen heading inland along the River Clyde. Video footage uploaded to Facebook and shared on several news stories showed two females and a juvenile. They were also seen spyhopping at the same time, perhaps trying to figure out where they were and how to get back out to sea.

Spyhopping is quite rare among orcas in Scotland. I have seen the Southern Residents spyhop, to look around and get their bearings, with eyesight far better than ours. What a treat it would have been for passers-by to see that in the Clyde. A slightly longer video was uploaded to *Dailymotion* which shows the same image of the two females and a juvenile with, trailing just behind, a mature male. Four of them. Another video clearly shows six animals, including a juvenile and a large bull.

Could this be the West Coast Community? Could they really have had a little one? I can't tell you how happy it would have made me to know that they had been able to reproduce after all. Maybe all was not lost.

It isn't unusual for sightings to be made a little further south (and fully marine) near the Isle of Arran, but orcas are rarely, if ever, spotted heading into the River Clyde. They were captured on camera between Dunoon and Gourock, near the head of Holy Loch and Loch Long.

In an article published in the *Telegraph*, it was said that this was an unusual sighting as orca are usually found in the cold waters of Norway and Iceland. It went on to say that orca in the UK are usually seen in Shetland and the Orkney Islands, but in no article was there any mention of the West Coast Community. Sadly, despite being our resident pod in the Hebrides, it appears that still few know about them.

The male is the easiest to identify. The dorsal fin was large and fully upright, so clearly it wasn't Floppy Fin. There wasn't a large nick out of the rear near the base so it wasn't John Coe either. Could it have been Aquarius or Comet? No, and it was most likely that this group was a pod from the Northern Isles who split their time between Shetland and Orkney, or some offshore orcas of which we know little, that occasionally come close enough to the coast to be seen.

When it was a sighting of just four I did get hopeful, but as the video of six was released, including a juvenile, even though I hoped it was our West Coast Community, I knew the chances were slim to none. They are simply never seen together in a big group.

So Comet is not the first, nor will he be the last, orca to pay a visit upriver. However, Comet's days are at best numbered, and most likely he has already passed on. He was estimated to be twenty feet long in 1977 and photos show his dorsal fin had already sprouted. Scientists think that he would have been at least nineteen years old in 1977. This would mean he was born in 1958, the same year as The Artist Formerly Known as Prince and Michael Jackson. If so, he has outlived both.

Taking a look back, in 1958 Harold Macmillan was Prime Minister of the UK and Eisenhower was President of the USA. There were 2.9 billion people on the planet; today there are more than 8 billion. If Comet was born in 1958, it means that by 2022 he would be at least sixty-four years old.

That is a long time, in fact longer than males are thought to live. While females have been known to survive well into their eighties (and Granny in the Pacific Northwest well over a hundred) males have a distinctly shorter life, with an average life span of thirty-one years and a maximum thought to be between forty and fifty. As we now know, John Coe was a mature bull when he was seen in the 1980s so is likely to be around sixty years old now, again older than the supposed maximum.

Based on his estimated age when he was known as Dopey, we know Comet would be double the average age that bulls live in the wild. Longevity such as this is testament to the value these animals place on social structure and community. If in old age individuals had nothing to offer, they would likely be disowned and left to fend for themselves.

As with humans, age brings ailments and a slower pace, but orcas do not seem to see this as a disadvantage and have been known to support the elderly, bringing salmon to them as offerings when they are too tired to hunt for themselves. The elders are the key to their culture, whether it is for babysitting the young or passing down knowledge and wisdom.

What benefit did Comet play in the group? There was no offspring to babysit, no next generation to pass down knowledge to or teach the family's cultural ways. He was presumably too slow to be of any real assistance in a hunt and potentially too old to breed, although in the Pacific Northwest it is thought some females seek out older and wiser males. He went on for many years, though, maintaining his bonds with John Coe and Aquarius. Perhaps, collectively, orcas are aware of their own demise and stay together for companionship until their own fateful day. Or perhaps there is more we need to learn. Perhaps sixty years really isn't that old and, against all odds, he is still out there.

We can only hope, even though we know that it is unlikely we will ever be blessed with seeing him again. The day I saw him, on 3 September 2014, he was hanging around with Aquarius, who was sighted many times in 2023. However, that day was the last confirmed sighting of Comet and the chances of seeing him again diminish by the day. This means that I was one of the very last people who saw, or will ever see, Comet.

Through researching and writing this book, I have learnt so much about this group of orcas, and my sense of compassion and empathy for them has grown immensely. Had I known then that our encounter would be the last anyone would have with him, I would have been overcome with a different set of emotions. Rather than being merely thrilled and exhilarated, I would have felt immense gratitude for both his extraordinarily long life and for the opportunity to see him and feel his presence. I would have felt a tremendous remorse for how we have failed him and his family.

How can it be that we have failed to protect these iconic creatures, a group so precious and unique to the British Isles? We could have, and should have, done more to understand them. It is a travesty that we have allowed them to slowly, one by one, pass on. Knowing he will likely never be seen again brings so much sadness to me and serves as a reminder that, before long, someone will have an encounter with John Coe or Aquarius and not know that it is the last time they will be seen. Every sighting of very probably the last two becomes even more poignant and precious, and is an experience to be truly cherished.

MOON

The circle of life

MOON (004) WAS FIRST RECORDED by the HWDT in 1992, the same year as John Coe, Floppy Fin and Nicola. As with John Coe and Floppy Fin, he was a mature male with a fully developed dorsal fin. His dorsal fin had a moon-shaped nick in it, hence his name. The last confirmed sighting of Moon alive was in 2001, but it is suspected by the HWDT that he died in 2008 in an orca stranding. If so, it would have meant he was at least twenty-eight years old, which is about the age that male orca are estimated to reach in the wild. It is quite possible that Moon died from old age. Male orcas have an average life expectancy of thirty years, but some will live into their sixties, as we have seen with John Coe and Comet.

Individuals within the West Coast Community are not seen each and every year. A few years can pass between sightings, so it is really hard to distinguish between when they are just not being seen and when they have died. With so many marine mammals swimming in our oceans, we rarely get to know about deaths: most just sink without a trace.

Orcas are the apex predator in the oceans. Nothing hunts them but mankind, so it is unlikely that Moon died due to predation. Just like other animals, including humans, as we age we get slower and less agile. Maybe he was unable to hunt his own food; even small seals and harbour porpoises require a certain level of agility, strength and stamina. However, as we already know, orca who feed on other marine mammals work in groups but share with other members of the family regardless of whether they were directly involved in the hunt.

Being social creatures with strong family ties, it is unlikely that John Coe and the others would have left Moon to fend for himself. That being said, it is not really known who he used to associate with and how much time he spent with them. He didn't seem to form the close male–female dyadic associations that the others did, such as Occasus and Aquarius, Nicola and Floppy Fin, Moneypenny and John Coe, and Puffin and Comet. Moon was more of a mystery, and he took his secrets with him.

When a whale dies, it will sometimes float for a while, owing to the gases that build in the stomach as the body decomposes. While buoyant and floating, the carcass can sometimes be washed up on the shore with an incoming current, where the stench from the rotting carcass is a good advertisement to scavengers that there is a free meal nearby. Shortly before I arrived on the west coast of Vancouver Island in 2016, the carcass of a young gray whale washed up off Florencia Bay, between Ucluelet and Tofino. The beach had been closed to the public over winter for road improvements, so the whale lay undiscovered by humans . . . but not undiscovered by other wildlife.

The carcass was soon surrounded by footprints in the sand from wolves, coyotes, birds and various other animals eager to get an easy dinner, and nothing was wasted. Whale skin is incredibly tough, so birds such as red-tailed hawks, ospreys and bald eagles will usually start to work away at the soft parts, notably the eyes and anus. If the mouth is agape, they can get inside and work from there. Sometimes the belly of a decomposing whale will become so distended with gases that, when the pressure rises too far, it simply explodes. Chunks of blubber can be blown down the beach, breaking parts of the whale up into more manageable pieces for scavengers. Bigger animals such as wolves and coyotes are able to strip the skin and get to the high-energy blubber. Maggots will finish anything left, and these processes eventually leave just the bone.

Whale bones are something of a collector's item in coastal communities. Left to bleach in the sun, they adorn front lawns and verandas of homes from Lerwick to the Isle of Lewis, and wherever in the world whale strandings are commonplace. It is not really the ghoulish practice it might seem; rather they are kept almost as a celebration of these animals and a fascination of everything about them.

Sometimes when a whale gets washed up on the beach, scientists learn about it before the body decomposes too much. In these cases, a necropsy will sometimes be carried out to determine the cause of death and to learn as much as possible about the species.

During my internship with the Vancouver Aquarium in 2016, I attended a training session on necropsies being held at the Pacific Biological Station in Nanaimo, on Vancouver Island, where a river otter was the subject of examination. This specimen had died three years previously and been in the freezer ever since. Due to the number of attendees, and it being a glorious day, the training was conducted outside, which was a blessing because the stench was pretty awful. However, when a large whale carcass is stranded on a beach, there is little chance that it will be moved so, often, the necropsy is done there and then.

Recalling this reminds me of a video I once watched of a fifty-tonne sperm whale that died on a beach in Taiwan in January 2004. It was decided that the best thing to do would be to transport the entire carcass back to the laboratory. A museum was also interested in keeping the skeleton for display.

A large flatbed truck was manoeuvred onto the beach and the whale lifted using a crane, slings and a whole lot of people. The

whale by this point was decomposing nicely and gases were building in the stomach. As the truck made its way from the beach to
the lab in downtown Taiwan, people stopped in their tracks to
stare and take pictures. It isn't every day that a dead sperm whale
is transported through your city.

Huge crowds gathered, filling the pavements. People left their
cars to get a better look. Inside the whale though, stomach gases,
mainly methane and hydrogen sulphide, steadily increased, and
pressure on the stomach wall built until ... BAM! ... the whale
exploded, sending blood and guts and chunks of blubber flying
down the street. Cars dripped red and blood ran down shop
windows. It was a horror scene.

Anyone who has done a necropsy will tell you that the stench
does not come out. Even if you are head to toe in 100 per cent
waterproof gear, you can bet your bottom dollar you will stink of
dead whale for some time to come. I guarantee that everyone
who lined the street that day can attest to that.

The intestines were scraped up and the body and organs
continued their journey to the university for analysis. It took a
team of sixty people ten days to complete the necropsy. The stench
of the body must have been indescribable by the end.

All of this amounts to the principal reason why necropsy on
large whales is usually done in the field. It is a case of logistics, but
working in the field is far from easy and can even be dangerous.
Tides can come in, meaning you might only have a short window
of time to work until the whale is covered in seawater again. It can
be slippery underfoot, and not just from the seawater: blood is
pretty slippery too. You are completely exposed to the elements,
and being too hot or too cold are real issues. It takes many hours
to conduct a full necropsy, approximately three hours for something small like a harbour porpoise but up to twelve for a large
whale for just the basic operation.

The bigger the whale, the bigger the team needed. These teams work fast and systematically – you do not want to get in their way. Starting with an external exam, they will take photographs of the body, the eyes, the teeth/baleen, fins, and any cuts, scrapes, signs of disease will be photographed and documented. The sex will be determined: females have mammary slits present, and a male is usually pretty easy to identify as the build-up of gases pushes the penis out from its internal cavity. Sex can also be confirmed from the internal exam.

I won't go into the full details of how the actual necropsy is done but, needless to say, samples are collected of each organ and tissue for thorough examination and an attempt to determine the cause of death. They are transported back to the laboratory for analysis, but it can take months before the results come back and a cause of death is provided. Sometimes, there is no obvious cause, and the rest of the carcass has to be disposed of. An untouched whale carcass is sometimes left on the beach if it is in a remote area, providing food for other animals. If the carcass is near an inhabited area, it will likely be dragged back out to sea when the conditions allow. If a carcass has been subjected to a necropsy the team has a responsibility to dispose of it, which usually means incinerating the remains.

While it might seem disrespectful to cut open these incredible animals, we can learn so much from them. Knowledge gained from necropsies has given great insights into the lives of many of the world's most elusive species, knowledge we would never have gained otherwise.

However, most carcasses just sink to the sea floor. These sunken bodies, known as whale fall, provide a vast amount of food in an otherwise barren landscape. Some carcasses will sink to the deep ocean basins, where they will feed an entire community of animals for decades. Animals such as starfish, brittle stars, hagfish and eels will feast upon these rare opportunities, leaving the bones completely clean.

These sunken whales can act as stepping stones between one community of invertebrates and another. With the deep seafloor so barren, organisms are often hostage to one feeding spot but when a whale carcass falls the scent will spread for miles, encouraging other creatures to seek out this new food supply. Once there, the journey to another feeding spot might not seem so far away.

There are three stages to a whale fall. The first is the mobile scavenging stage, when the smell of the carcass is detected by animals such as hagfish, sleeper sharks and ratfish, who consume the soft fleshy parts.

The second phase is the enrichment, opportunistic phase where worms, crustaceans, molluscs and other small invertebrates feast on the remaining blubber until all that remains is the skeleton. They will also dig into the sediment surrounding the whale to get right underneath, where they can consume the now heavily organically laden sediment itself. It can take a further two years to strip the animal back to the bones. Huge numbers of organisms will live on the carcass during this period, as many as 40,000 in every square metre. That is incredible in what would have been an otherwise lifeless patch of sand or mud.

The final stage is the sulfophilic stage, which can take decades. Chemoautotrophic bacteria begin their feast on the bones, breaking down the lipids contained within. Doing this, they produce sulphur, which attracts a whole range of organisms such as mussels, worms and snails. The community that gathers here can last for many years and is thought to be the largest community on the deep ocean floor.

Whale falls are an incredibly important part of deep-sea ecosystems and can sustain an entire food web for decades. Trawling sometimes brings up the remains, and it has been found that, in the whale skulls trawled up from the seafloor off Scotland, a specialised species of mytilid called *Adipicola simpsoni* lives in huge abundances within the skull.

If Moon's body sank to the seafloor, his death will have brought life to many other animals. However, that isn't to say that his body couldn't have washed up somewhere already but was not found or was too decomposed to identify. There are countless remote and uninhabited beaches and rocky outcrops around the Hebrides, meaning it is quite possible that he stranded but has not been found.

The Scottish Marine Animal Stranding Scheme runs a database of all marine-stranding reports. There have been eight orca strandings in the Hebrides recorded between 1995 and 2021, with fourteen in total in Scotland, including Shetland. From these eight, three were found to be female, two confirmed as male and a further three were too decomposed to identify.

One of these was an eight-metre-long male that washed up in 2008 at Sandwood Bay in north-west Scotland. Sadly, it was so badly decomposed that a full necropsy wasn't possible, nor was it possible to make a positive identification of the individual. The cause of death was unknown as there were no obvious signs of trauma or entanglement. However, many believe that this was Moon due to external clues such as the position of the eye patch and his large size.

If this was indeed him, two interesting discoveries were made. First, he had minke whale baleen in his stomach, confirming that the West Coast Community hunt minke whales and not just porpoises. Second, genetic analysis showed that this individual was closely related to the Type A Antarctica orca, and more distantly related to the other orca (Type 1) that we see around Scotland. That is to say that it looks like Moon, if indeed it was Moon, was part of a splinter group from Antarctica.

There was another stranding in January 2015 of a male orca that washed up on a beach, this time near West Gerinish on South Uist. This would have been fourteen years after the last confirmed sighting of Moon, but it is not impossible that this could have been him. He was badly decomposed, too decomposed for a proper necropsy, so the team at the National Museum of Scotland (NMS) made the journey from their base in Edinburgh to collect the body and bring it back for preservation.

The NMS has a huge collection of marine-mammal skeletons and specimens, so good that many international researchers are attracted there to obtain data and make comparisons against species found in their own country. Dr Kitchener, the principal curator, was lacking a good orca skeleton, so this find was just too good to miss out on.

The original report was of a whale about three metres long. It was so badly decomposed that it wasn't even reported as being an orca and the size was wildly out. It took a few days to get plans in place and secure funding from the SeaWorld and Busch Gardens Conservation Fund. With that, on 3 February, two of Dr Kitchener's team, Georg Hantke and Alan Lothian, set out from Edinburgh. Upon arrival in South Uist, they found that the male was actually 7.8 metres long, much bigger than the three metres reported. This was going to take the pair longer than anticipated.

The body had washed up on a public rocky beach right next door to Rangehead, a Ministry of Defence base for launching missiles. This turned out to be pretty handy, as a few members of staff from the MoD missiles safety team, including their explosives manager, stepped in to help Georg and Alan collect the body. It was a grim, stinky and tricky task, taking over twenty hours to measure the animal, extract tissue samples and strip off as much flesh as possible before dividing the skeleton into manageable sizes for transport back to Edinburgh. A van ride, a ferry and another van ride later, the carcass was unloaded and the process of preserving the skeleton began.

Any remaining excess flesh had to be cut away and the bones placed in a salty bath to help remove any grease from the blubber before being heated to 37°C for several weeks to strip off the last tiny bits of flesh. Once the skeleton was laid out and cleaned, it became evident that this male had a severely broken lower jaw. There were no signs of healing to the bone, so it is likely that this incident happened a short time before the death or even after the death, but likely not during stranding.

Perhaps this old male had a collision with a boat which caused it to die? Or maybe it was attacked by another animal shortly after death, resulting in a broken jaw. I interviewed the very gentle and calm Dr Andrew Kitchener and he revealed that the jaw was actually broken on both sides and, furthermore, several ribs were also broken. He tentatively concluded that this individual died from a collision with a boat. Since the bones did not show any signs of healing it is likely he died on impact or shortly afterwards.

Speculation had arisen when the body was first seen by the team. The size of this individual was much larger than the typical size of the more commonly seen Type 1 orca. In fact, its size was much closer to what we expect from a Type 2 or West Coast Community individual. Interesting, as well, is that the teeth were sharp and not worn out. The Atlantic Type 1 orcas that come over from Iceland are generalist feeders on fish and seals and typically have very worn teeth, thought to result from sucking up tons of tiny herring and mackerel. This individual had sharp teeth like the ones we see in the West Coast Community and the Type 2 killer whales. Could he be Moon?

There hasn't been a confirmed sighting of Moon since 2001 and the decomposed body found in 2008, which has been speculated and even assumed to be him. It is quite possible that he died from old age, but this more recent stranding in 2015 also seems to fit his description. Could this have been Moon and he had been around between 2001 and 2015 but avoided the cameras? Maybe there are other members of the West Coast Community that we just don't know about. In the 1980s it was thought that there were at least twenty, so it is not beyond the realm of possibility that there are more; unlikely but not impossible. To this day it remains a mystery as to which one of these stranded whales was Moon, or if indeed, either of them were. Perhaps he is still out there.

LULU

The world's most polluted whale

LULU (006) WAS ADDED TO THE West Coast Community catalogue in 1995, a few years after the initial group in 1992 of John
Coe, Floppy Fin, Nicola, Comet and Moon. That isn't to say that
she wasn't around in 1992, just that she was not seen by HWDT
until then. While the others have names that are more descriptive, such as Floppy Fin with his collapsed fin, Moon with his
crescent shape dorsal fin and Nicola with that distinctive nick in
her dorsal fin, Lulu's name is just for fun and a nice tribute to
someone who worked to study and protect them. She was named
after Alison Gill, who worked with Sea Life Surveys when this
community of whales was first being catalogued. Alison had the
nickname 'Lulu' and her teammates suggested she deserved to
have one of the whales called after her as she was working so
hard on the project.

When she was first spotted, she was already a mature female,
likely to have been born sometime before 1980. The sparsity of
orca sightings means it is hard to know when exactly they were
born, where they go or who they hang out with.

Like John Coe and Comet, she appears to have been a big
roamer, and was sighted in 2009 off County Mayo on the west
coast of Ireland. So, while the West Coast Community are 'resident' to the Hebrides, we know that nearly all have made the
journey south to Ireland at least once. She was not alone but
with Floppy Fin, Nicola, Occasus and Aquarius. She seems to be

fairly sociable as, on every occasion that she has been sighted, she was with at least two others from the West Coast Community, and most often in a group of four or five.

On every sighting, Aquarius was seen with her. As we know though, Aquarius isn't always with Lulu as, when I saw him, he was with just Comet. However, Comet is another whale she is often sighted with. Which makes me wonder, perhaps she wasn't that far away when I saw Comet and Aquarius on that cold day in 2014. Maybe she was just around the headland? I will never know. In 2014 she was seen swimming around the beautiful coastline, but we didn't have another sighting of her until 2016. As is often the case with the West Coast Community, you can see them one year, but not again for a few years. There is a lot of ocean for them to explore.

While many orcas appear to have their go-to male-female partnering, the bonds in the group are likely to be relatively fluid, with their associations changing, perhaps for different tasks. Maybe some like to hunt together, but to play and relax with others. She has also been seen with each of the other females in the group, so seems to keep a balanced company of males and females.

While females are harder to tell apart than the distinctive males with their enormous dorsal fins, Lulu was distinguishable from the others as the tip of her dorsal fin was a little more pointed. That said, it still takes a trained eye to identify the females. She also had a distinctive mark on her saddle patch at the base of her dorsal fin which is unique to her. These small differences are important in identifying who is who, especially in times when one life comes into the world, and when one life leaves. As was the sad day in 2016 when Lulu was last seen, this time alone and dead on the beach.

★   ★   ★

Lulu hit the headlines in early January 2016 with the devastating discovery of her body washed up on Crossapol, a rocky beach on the Isle of Tiree, the most westerly of the Inner Hebridean Islands.

The cause of her death was obvious. She had severe lacerations along the base of her tail fluke and the markings of fishing lines which had been so heavy on her tail that they left permanent imprints. Looking at the markings and the way they cut through her skin, experts suggested that she likely had a lot of heavy fishing line and rope wrapped around her tail. There were no ropes or nets on her when she was found, strangely. This is the case for about 70 per cent of dead stranded cases. However, based on markings left on minke whales that have been entangled, it is thought that she had become caught up in crab-fishing creel lines.

This is, sadly, all too common for minke whales, but not something that we are used to seeing with the more agile orca. The problem with creel lines is that they have buoys and heavy crab pots attached to each line of synthetic rope, and they are very long, sometimes several kilometres, meaning they would have been incredibly heavy. The extra drag created from all this weight would have meant her swimming would have been severely hampered and she would not have been able to hunt properly. It is likely that she died a long, slow death from starvation and exhaustion.

The necropsy that followed confirmed that her stomach and intestines were empty. She hadn't eaten for at least several days. It also confirmed that she had swallowed a lot of water, a sign that she was struggling to stay afloat as she became exhausted, gulping in mouthfuls of water in her desperation. What an incredibly sad end for such an amazing creature.

On the *Silurian*, aside from noting all sightings of marine mammals, we also had to record each sighting of a creel buoy. I remember that, at the time I and, I think I am right in saying, most of the other crew members, felt frustrated having to call out. Estimating the distance between the boat and each one was not easy and, when it was located in heavily fished areas, seemed to fill the whole thirty-minute slot. Calling out, time and time again, 'Creel buoy, fifty metres, starboard'.

The creels are designed to trap lobsters, brown crabs and velvet swimming crabs, and the buoy simply marks the start of lines which can have twenty to a hundred steel-framed creels attached by a short rope. The main line is usually one kilometre long but can be up to four kilometres. There are thought to be over a thousand creel fishing boats operating in Scotland, so creel fishing is big business.

We were informed at the start of the trip that creel fishing is particularly dangerous to marine mammals, particularly minke whales, of whom, each year, about twelve die in the Hebrides. Of these, five or six are thought to have died following entanglement in creel lines.

Although we knew the risks the creels pose to marine life, it just didn't feel that important to note all of them. The fishing industry in the Hebrides is a major source of income and a way of life. Was it really going to make a difference, did they really pose that great a risk in the grand scheme, and would the data we were gathering even be used? We wondered, but when the news of Lulu came in, and I saw the photo of her bloated body on the beach with horrific lacerations, so clearly from the fishing line, I understood that the data we were gathering was important.

Without monitoring these things, how can we possibly know how prevalent this type of fishing is and the areas in which it is particularly intensive? If we know where there are hotspots for marine life and if they coincide with areas of intensive crab and lobster fisheries, we can work to minimise the impact. I understood that day, on a level I hadn't previously, that we have a devastating impact on our marine life and must try harder to live harmoniously with nature.

Thankfully, in 2018 a new initiative called the Scottish Entanglement Alliance (SEA) was established, aiming to understand the impacts and risks of entanglement in creel lines in Scottish waters. It is a joint effort from HWDT, NatureScot (formally Scottish Natural Heritage), Scottish Marine Animal Stranding Scheme (SMASS), Scottish Creel Fishermen's Federations (SCFF), Whale and Dolphin Conservation (WDC) and the British Divers Marine Life Rescue (BDMLR). These organisations work closely with the Scottish inshore fishing industry to work towards building strategies to help reduce the threat of entanglement.

Recent research by this group has discerned that the problem with creel lines and entanglement is actually worse than previously feared. However, with co-operation from the fishing industry, new ropeless systems have been trialled with success and the hope is that they will be used widely. If so, this is an incredible development. Sadly, it came too late to save Lulu. We will never know how she got entangled or fully understand how horrid her last few days were, but we can imagine by listening to stories of other entanglements that have been witnessed.

In April 2016, a few months after her death, an orca was seen in creel lines off the California coast in the Monterey Bay National Marine Sanctuary, where a group had been feeding on a gray whale calf they had killed that morning. At that time of year, gray whale mothers and their new calves head from their winter breeding grounds in Mexico's Baja peninsula to their summer feeding grounds off Alaska. These young calves are such easy prey for the orca that many new mothers lose their young ones and have to make the rest of their journey on their own.

Taking it in turns to feed, the orcas were moving in and out of proximity to the carcass, where there were also several sets of crab-fishing gear with pots, creel lines and buoys marking the crab-pot locations.

A male orca swam alongside one of the buoys and got its line wrapped over his body and dorsal fin. It took him four dives to free himself but, thankfully, on the last attempt he was successful. It could easily have turned out a different way. This entanglement was witnessed by the Whale Entanglement Team at Marine Life Studies, a not-for-profit organisation that specialises in dealing with whale entanglements. They reported that it wasn't clear if it was purely accidental or if the orca was playing with the creel lines, being the inquisitive and playful creatures that they are. Sadly, despite being so agile and intelligent, innocent play can some-times get orcas into trouble, leading to dire outcomes. This time the orca managed to free himself and swim back to the safety of his family, but it raises the question: was Lulu just playing a game with the buoys and got herself entangled, or had she simply not seen the thin creel lines and accidently swam into them?

While it is usually rare for these agile creatures to get entan-gled, it is still all too frequent an occurrence. On 6 September 2021, a well-known and much-loved Scottish orca, number 151 from the group 27s, was found dead on the Orkney island of Papa Westray. According to the Scottish Killer Whale Catalogue 2021, the 27s are a group of eight orcas who belong to the Northern Isles population that spends most of its time between the Shetlands and Orkney. This family has also been seen off the Faroe Islands but has never been recorded in Norwegian waters. It has also been seen each month of the year in Shetland, so seems to be at least semi-resident in Scotland, feeding on seals and porpoises.

Orca number 27 is thought to be the matriarch, having had five children. Her first, 034, was born about twenty-five years ago and her last, 150, around 2015. Orca 27 was thought to be about forty years old and heading towards the menopause. She had already taken on the role of doting grandma when her third-born, female 073, gave birth to number 151 in late 2015 or 2016 and then number 153 in the summer of 2019. As we now know, 151 has since perished, bringing the group size back down to seven, including just one baby.

The death of 151 was caused by entanglement. It had rope markings around the tailstock and, while we can't be sure whether this was from active fishing gear or discarded 'ghost' fishing gear, the result was the same. The whale became entangled and couldn't get to the surface to breathe and so suffocated.

Aside from the rope markings, this whale was thought to be in good condition, nice and healthy, feeding well, with no evidence of plastic debris in its stomach. We can take as positives from this event that, at least health-wise, it appears this family is doing well. But death by entanglement is such a sad end for a young individual of just six years of age.

The death of Lulu was a devastating blow to all those who had been tracking her movements in the years since she was first identified as, indeed, it was for all marine-mammal lovers. Her death reduced her group to just eight individuals, including only four females. While any death is devastating, losing a female means the loss of any possible new calves, and many viewed Lulu's death as the final nail in the coffin of the West Coast Community. The group was now too small and too old to bring any new life into the world. The silver lining, if it can be called that, came from the necropsy that followed.

A team of scientists from the Scottish Marine Animal Stranding Scheme (SMASS) headed straight to the Isle of Tiree when they got the call that an orca was stranded on the beach. The SMASS respond to stranding reports that come in from all across Scotland on marine mammals, turtles and basking sharks and has been in operation since 1992. The data they collate is crucial to understanding patterns in mortalities and the health of populations and the wider ecology. The team is headed by Dr Andrew Brownlow, an expert in marine mammals and a leading veterinary pathologist.

By the time Lulu was discovered, it was estimated that she had already been dead for four days, which meant that her internal organs had already begun to break down and decompose, making it impossible to do a full necropsy. However, they were able to state that there was no clear evidence of significant disease, making her death from entanglement the most likely cause.

As already described, she had the hallmark markings of entanglement from creel lines that had evidently been wrapped around her tail fluke. She also had abrasions on her pectoral fins, but these could have occurred after her death as she was washed onto the shoreline. Despite her partially decomposed state, the SMASS team learnt a lot and took samples for subsequent laboratory testing.

Several different teams were involved in the laboratory work. Dr Andy Foote, the leading expert on this family, looked into the genetic analysis of the individual, and Paul Jepson from the Zoological Society, London, conducted a toxicity analysis.

As we saw in John Coe's chapter, there are thought to be ten different ecotypes of orca and two that form the orcas of the Atlantic. It is suspected that there are also offshore orcas and not all populations necessarily fit into these two ecotypes. I know, it is confusing, but our understanding is ever evolving.

The Type 1s are on the smaller side, have a horizontal eye patch and have worn-down teeth. This includes the prey-switching herring-seal-eating ones from Iceland, who are summer visitors to Shetland, and the Northern Isles community which are at least semi-resident to Shetland and Orkney and feed on small marine mammals.

The Type 2s are much larger, with a slanted eye patch and sharp teeth. The stranding thought to be Moon in 2008 added to the theory that the West Coast Community are the Type 2s and different from other orcas in Scotland, though without genetic sampling this can't be confirmed.

Upon inspection, she was found to have the characteristic sharp teeth of Type 2 Eastern North Atlantic killer whales. So much can be learnt from teeth; not just the ecotype but also the animal's age can be estimated. Much like the rings of a tree, a growth layer of dentine is created each year of an animal's life. Counts of Growth Layer Groups (GLGs) is the term used to describe this process, and from this the age of the marine mammal is estimated.

When the necropsy team examined the teeth, they confirmed that Lulu was at least twenty years old. Confirmed, as Lulu was first identified in 1995 and was at least twenty-one then, more likely in her thirties as she was not a calf or juvenile when first catalogued.

Dr Foote identified her as Lulu by comparing her features to the HWDT catalogue developed using the many community sightings made by the dedicated public and researchers, based on her slanting eye patch, size, teeth, dorsal-fin shape and the unique marking on her saddle patch.

Dr Foote specialises in genetic analysis, investigating how species can separate and evolve into multiple species. In fact, the classification of Type 1 and Type 2 Eastern North Atlantic killer whales was a brainchild of his, so there is no one better qualified to look into Lulu's genetics. The results, however, were unclear.

The match for a Type 2 wasn't perfect but it appears that Lulu is possibly a descendant from an Antarctic population, though the jury is still out on that. His research also used microsatellite genotyping which, in basic terms, is where short strands of repeated DNA (the microsatellite) are used to determine the genetic make-up of an individual and can be used to see what genes were inherited from their parents (genotyping). The results showed that she was highly homozygous, meaning she had the same genes inherited from both her mother and father.

This matters because it indicates that Lulu was inbred, at a level that is unprecedented in any other orca population. As we discussed in Nicola's chapter, the Resident orcas of the Pacific Northwest have distinct calls for their family group, which allow them to determine who they are related to and therefore who they should avoid breeding with.

Sadly, it looks like Lulu comes from a highly inbred family, possibly owing to there having been no other options. With so few males around, it is a case of mate with your brother or don't mate at all. However, even more interesting is that the necropsy revealed from her ovaries that Lulu had never been pregnant. She may have tried to mate with John Coe or Aquarius or any of the other males, but just not fallen pregnant. Quite likely she was so inbred that she couldn't.

However, the question of why she has never been pregnant, and perhaps why the family as a whole has never been seen with a calf, if not definitively answered by their level of inbreeding, was also addressed through toxicity testing of her blubber.

When the Zoological Society of London examined Lulu's blood and tissue samples they made a rather shocking discovery. They found that she had an exceptionally high level of a toxic compound called PCB in her tissues. In fact, it is quite possible that, of all marine mammals ever tested, Lulu had the highest concentration.

PCB stands for polychlorinated biphenyl, an organic chlorine-based compound, a chemical that was once widely used in coolants in electrical equipment, but also in plastic products, paints and rubber products. Basically, PCBs were used in the manufacture of a whole load of everyday products. However, before long, it was found that PCBs are actually pretty harmful. They were proven to be carcinogenic, that is, they cause cancer. They are also known to affect the endocrine system, most notably impacting thyroid function.

They are immunosuppressant, meaning that after exposure you are likely to get ill a lot more and ... here is the big one for Lulu ... they are also known to affect fertility. In a word, PCBs are evil. Because of this, they were banned in the 1980s, but it appears that they are persistent little buggers that stay in the environment for decades. They are termed Persistent Organic Pollutants, which means that, even though they are no longer in use in the UK, we still find them in the environment and, as we saw with Lulu, in shockingly high concentrations.

Orcas are particularly affected by PCBs. In 2000, a study was conducted to compare the PCB levels in the blubber of resident and transient orcas in the Pacific Northwest. It was found that levels in the transients (remember, orcas who eat other marine mammals rather than just fish) were much higher than those in the residents. This was not altogether unexpected as PCBs get stored in the tissue of animals. They enter the food chain when caught by tiny filter-feeding plankton or small filter-feeding molluscs. The concentration at this point is miniscule. Tiny fish eat the plankton, and they eat lots of it, so small amounts of PCBs accumulate in the tissue of the fish. Bigger fish eat a lot of the little fish, bioaccumulating more PCBs. Resident orcas consume the big fish, such as salmon. Over a lifetime of eating toxin-laden salmon, PCBs build up in the blubber of the resident orcas. The situation is worse for the transients as there is another layer to the food chain. Other marine mammals, such as seals and porpoises or larger whales, eat the fish and store PCBs in their body.

While experts expected the transients to have higher concentrations than the fish-eating residents, it transpired that the transients had much higher PCBs than anticipated. The levels were so high that it is thought that PCBs were a great risk to the orcas.

The story is bleaker still for orcas on our side of the planet. Another study was conducted in 2016 comparing the PCB levels from marine mammals across Europe. It showed that marine mammals in general, not just orcas, in the north-east Atlantic have the highest level of PCB toxins anywhere in the world, including the transient orcas off Vancouver Island. The average orca in the north-east Atlantic has about 150mg/kg of PCB in their blubber.

Even before Lulu's necropsy results, it was assumed that our West Coast Community was likely to have pretty high levels of PCBs in their blubber as they eat other mammals. It was also thought that, given that they live in a pretty unpolluted part of the world, it might not be so bad. Furthermore, since they all seemed to be pretty old it might indicate that the PCB load wasn't too high. They don't seem to succumb to minor infections, suggesting their immune system hasn't been weakened too much. They are living a long time and not dying young from cancer or thyroid issues. So there was hope that our West Coast Community might be spared.

Male orcas usually have higher PCB concentrations than females because they can't get rid of it. A female will pass on some of her contaminant load (up to 60 per cent) to any offspring she is carrying and then also in her milk when she is nursing. This is known as mother-calf-transfer. If she has been able to reproduce, her PCB load reduces; however, this obviously isn't good news for her baby. It is particularly bad for the first offspring, who receives fifteen or more years of the mother's toxic load in her milk, leading to high infant mortality rates. Subsequent calves get on average just four or five years' worth. The males can't offload like this, so the level just keeps getting higher until they are so immunosuppressed they tend to die younger from smaller infections.

When the results came in, it was pretty shocking. Not only did Lulu have a high concentration of PCBs compared to other marine mammals in the Atlantic and the transient orcas of the Pacific Northwest, but she also had the highest load ever recorded and quickly became known as the most polluted whale on the planet.

Poor Lulu. She had a concentration of PCBs of 957mg/kg lipid weight in her blubber. For reference, the concentration of the PCBs in transient orcas is 251mg/kg lipid weight (for males) or 58.8mg/kg lipid weight (for females).

We thought that transient orcas in the Pacific Northwest were heavily laden, but they carry little in comparison to Lulu. The concentration of PCBs that is thought to damage the health of marine mammals is 9mg/kg, and the highest toxicity threshold, after which fertility problems can arise, is 41mg/kg lipid weight. Therefore, Lulu had nearly twenty-four times the amount of toxins in her blubber than is deemed safe.

It is looking ever more likely that we have found the reason why our West Coast Community of orcas has never been seen with offspring. They are so heavily polluted from the chemicals that mankind has created and mass-produced, that we have prevented one of our most iconic species from not only producing viable offspring but even falling pregnant in the first place.

With the Southern Resident orcas of the Pacific Northwest, we are currently witnessing the heartbreaking situation of incredibly high infant-mortality rates. Almost every calf that is born, sadly, dies just days or weeks after birth, if it is not stillborn. Back in July 2018, Tahlequah, or J35, gave birth to a calf that, just a few days later, died. The world was held captivated and heart-broken, watching helplessly as she carried her dead baby around for a staggering seventeen days and over a thousand miles before finally letting it go.

I saw a picture of just this behaviour when I was working at the Vancouver Aquarium a few years before. The photo had been taken from the air of a mama orca keeping her lifeless baby at the surface using the tip of her rostrum, the snout-like projection on the head, and it brought tears to my eyes. When we are taught in science not to attribute human characteristics to animals, told that emotions such as grief and sorrow are only present in higher-order animals such as ourselves, it makes me so mad.

Why do we think we are so special that we are the only ones to experience grief? If you look at those photos, really look with your heart open, you can't help but understand that we are far more alike than we previously thought. This tragedy of babies dying one after another is devastating for the Southern Residents. They so desperately need their numbers to increase, and we need the joy of knowing that at least one baby makes it through the tough first year. The West Coast Community, being so badly poisoned that they can't even conceive, are perhaps at least spared this heartache.

The other problem with PCBs is that, in a healthy individual, they are stored in the blubber. However, if there isn't much food around and the orcas begin to starve, or they get sick and can't hunt effectively, they metabolise their fat stores for energy. The same thing happens to us. In fact many of us do this intentionally by following the keto diet (or Atkins diet) where a high-protein/

low-carb diet is followed to encourage the use of our fat stores to help us look slimmer. When an orca enter ketosis (or starvation) it starts to release the stored PCB load into its bloodstream, making it more susceptible to illness and toxic-related side effects.

The Southern Residents are currently starving. There isn't enough Chinook salmon to keep them healthy, and drone photos have shown that there are now several individuals from this population that have 'peanut head', a term used to describe how the head starts to look when the fat stores are so depleted that the animal is near the point of death. Stored PCBs are then released into the blood, and this might be the reason for their very high infant-mortality rates of recent years.

Lulu didn't look emaciated though, and the necropsy revealed the cause of death as entanglement, not cancer or infection. How Lulu was able to survive so long and appear healthy, aside from infertility, when other populations seem to be affected more with lower concentrations is yet another unanswered question from the West Coast Community. Her skeleton and teeth have been preserved at the National Museum of Scotland in Edinburgh, along with the male skeleton retrieved in 2015 that we discussed in Moon's chapter.

The impact that PCBs and damaging fishing practices had on Lulu should serve as a warning. Today, we are paying the price for the environmental mistakes of previous generations, and we must make sure that we are not leaving mistakes of our own that our children and grandchildren will pay for. If we continue burning fossil fuels, cutting down rainforests and using toxic chemicals we will exacerbate the degradation of the environment as we know it.

We live, in general, as a society where convenience is put at a higher premium than nature, with fast food and fast fashion taking precedence over slow living and simple pleasures. Jane Goodall once said, 'We seem to have lost the wisdom of the indigenous people, which dictated that in any major decision, the first consideration was; 'How will this decision we're making today affect our people in the future?'

FLOPPY FIN

Nature versus nurture

FLOPPY FIN (002) IS A MALE ORCA in the West Coast Community with a very distinctive dorsal fin that 'flops' to the left. Along with John Coe, Nicola and Moon, he was first catalogued in 1992 and has been seen around the Hebrides many times. The last official record was back in 2011, so it was presumed that he had passed on until, in early September 2021, a decade after his last encounter, a sighting was reported on the community sightings platform run by HWDT, Whale Track, of three orcas near Loch Pooltiel, north-west Skye.

The whales were reported as being John Coe and Aquarius, nothing unusual there as they have been seen frequently in 2021, but the third whale was said to have a flopped-over dorsal fin. Could this be our Floppy Fin of the West Coast Community? Sadly, there is no photographic evidence so we can't be sure.

There is another orca known as 013 in the Scottish Killer Whale Catalogue that frequents Shetland more than the Hebrides. His dorsal fin flops to the right, whereas Floppy Fin's flops to the left. If it was 013 then this is also incredible as it would be the first time the West Coast Community has been seen to mix with orcas outside of their family. Without that photo, sadly, we will never know who it was. It is nice to think though, that it was Floppy Fin and that he is still around, as that would give us hope others are also out there, keeping a low profile.

Dorsal fins in orcas are used to identify individuals. The dorsal fin varies hugely between cetaceans. Some have just tiny fins, such as the humpback whale (*Megaptera novaengliae*) and the sperm whale (*Physeter macrocephalus*) while others have no dorsal fin at all, such as the beluga whale (*Delphinapterus leucas*) and the northern right whale dolphin (*Lissodelphis borealis*). The shape, size and position (or lack) of dorsal fins are some of the easiest indicators of which species you are looking at. Other factors, such as the size of the individual, colouration, blow height, shape and behaviour are also useful.

When you meet whales in the wild, quite often all you will see or hear is them breathing, displayed by a big mist, or blow of air, followed by the dorsal fin. You might even smell their fishy breath. Sometimes you get just a split second to take it all in and make an identification, which can be extremely challenging. Even the most expert marine-mammal surveyors, who spend months at a time at sea and whose job it is to identify whales and dolphins, will struggle with many species.

However, identifying orcas can be a relatively easy task because there is a huge difference between the male and female dorsal fins. Mature females and young males are pretty similar, triangular in shape with a curved posterior and a rounded-to-pointed tip. During adolescence, the male dorsal fin sprouts up with supportive collagen holding it in place.

The dorsal fin acts like the keel of a boat to give the animal stability in the water, preventing it from rolling over unintentionally. It also serves as a site of thermoregulation. Being warm-blooded, orcas must maintain their body temperature at around 97–99°F (36.1–37.2°C). Like other marine mammals, they have faster metabolisms than land mammals and, despite their preference for cold water, can overheat. The large dorsal fin is full of blood vessels that allow heat to escape into the air when they surface. It is also possible that, given the large difference in male and female orcas, the large dorsal fin of mature males acts as an advertisement for their fertility and prowess, much like a lion's mane or the antlers on male red deer in Scotland.

In the 1970s, Dr Michael Bigg made the discovery that dorsal fins of orcas in British Columbia are unique. In fact, all orca dorsal fins are unique, globally, with slightly different shapes, sizes and notches, much as our fingerprints are unique and used for identification. Additionally, the white-greyish patch at the base of each individual, the saddle patch, is also unique.

This discovery meant that every single orca could be identified, allowing associations between individuals to be established and individuals traced throughout their life. This knowledge revolutionised research on this species (and others) and has been the basis of all genetic and behavioural studies to date. It is simply fantastic that this simple observation has enabled decades of researchers to learn so much and arm the rest of us in the fight for orca conservation. It is this photo-identification of dorsal fins that has allowed us to recognise and track our ten from the West Coast Community.

A collapsing or collapsed dorsal fin is a fin that is or has begun to hang to one side of the animal's body, or that is no longer upright. It may also have some form of malformation such as a bend or wave, or a combination of both hanging and bending. A full floppy fin is quite rare and, in the wild, is thought to be due to a genetic disorder. Most stay upright, despite their enormous size, up to two metres in height, taller than a full-grown man, with only collagen to stiffen them. This is due to strong muscles.

Near constant swimming in a weightless environment means that the heavy fin is supported and grows straight. However, there are orcas in other parts of the world who have completely floppy fins, and others where the sheer weight proves challenging, with the fin in various stages of collapse.

Orcas are often given two names. One has a scientific basis, usually describing which pod it is from, offspring of which mother or catalogue number, such Floppy Fin's 002. They are often also given a more descriptive name, an aide-memoire, such as Floppy Fin, Moon, Nicola, etc. These names often stem from observations that set them apart from others, such as Nicola, with the large nick on her dorsal fin, and Moon, whose dorsal fin is like the crescent of a moon. A collapsed or collapsing dorsal fin is often the inspiration for their common name.

Always of interest, often amusing, names given to males with floppy fins from around the world include:

New Zealand population

Corkscrew. Also known as NZ15. He has a dorsal fin that is bent into a loose concertina shape with a wavy trailing edge. The whole fin leans to the right. However, the fin stays ridged when surfacing and does not wobble.

Captain Hook. Also known as NZ66. The top third of this dorsal fin is bent into a hook shape and the trailing end has a wavy appearance.

Wavey. Also known as NZ86. Wavy trailing edge to his dorsal fin. Unlike Corkscrew, the whole fin wobbles each time he surfaces. The wave and wobble became progressively worse with each year that he is sighted.

British Columbia

Hooker. Also known as B1, is from the Northern Resident population. He is from a family that is mostly made up of males. Born around 1951, he died back in 1998. He got his name from his dorsal fin, which hooks forward. Because he was so distinctive, he was one of the first ever orca identified by the Bigg's team.

Slingsby (B10, Northern Resident). A younger brother to Hooker, who also had a dorsal fin that was concertinaed and tilted to one side about 30 degrees. However it began to right itself in 2006.

Yaculta (B13, Northern Resident). Also a younger brother to Hooker. As with the other males in his family, he started out with a straight dorsal fin at sexual maturity but with time it started to curl over and by 2006 it had completely collapsed. He is one of several males in his family to have a floppy fin to some degree, adding strength to the argument that the condition is likely to be genetic in the wild.

Norway

Flappy (0063). He has a completely collapsed dorsal fin that flops to the left.

In 1993, Dr Ingrid Visser compiled a catalogue of all the orcas in New Zealand, which showed that seven of the males had abnormal dorsal fins. Of the total population, 23 per cent of adult males showed some form of collapsing or twist. Visser reported that this percentage is higher than in the British Columbia resident population, where 4.7 per cent were reported to have abnormal fins, and 0.57 per cent for the Norwegian population. However, only 0.1 per cent of the New Zealand population had what could be classed as a collapsed fin, similar to those seen in captivity.

Dr Bigg remarks that abnormalities develop structural weaknesses which eventually result in the curling of the tip, less rigidity and sometimes complete collapse. Therefore, it was thought that abnormal fins were due to the age of individuals.

Dr Visser postulated that abnormal fins in some cases might be related to events that had resulted in other issues such as body scarring. These could include an attack from another orca, or perhaps the stress of the attack caused the fin to start collapsing, but it couldn't be determined whether the two were related. She also noted that, while about 25 per cent of the population has some kind of dorsal fin abnormality, it is pretty rare for there to be complete collapse.

It is not really known if deformed dorsal fins are a disadvantage to orcas. Usually in nature a deformity in an individual means it is less attractive to potential mates, being less capable of surviving in the wild and slowing down a group. Therefore, it is often the case that individuals with deformities or disabilities are rejected, either at birth or sometime afterwards. However, stories are emerging where this is not the case, and not just for orcas but for other intelligent and social animals.

An orca nicknamed Stumpy lives in Arctic Norway. She has a severed dorsal fin, and a severely deformed spine. It is thought that she was born with a curved spine and that the severed dorsal fin was the result of a collision with a boat, likely because she is unable to swim very fast. Due to her disabilities, she is slow and lethargic, her breathing rapid and shallow, with lungs that are not as fully functional as other adults. Back in 1996, Stumpy used to travel with her mother, but in 2003 her mother disappeared, presumed dead. Without her mother to care for her, it was thought that she too would perish. Slow and sluggish, she was unable to hunt and was reliant on her mother to bring her food.

However, when her mother died she was adopted by other groups and, in total, Stumpy has been associated with up to seventeen different pods, all thought to be of the same family line. She is often seen with families, with the babies slipstreaming their mothers to conserve energy, the big males on the outside in their protective manner, and poor Stumpy trailing behind. She has to pump her tail fluke hard just to keep up, despite the others reducing their speed to help her out. Due to her extra efforts, she likely has to eat more than any of the others and, since she is unable to catch anything herself, the compassion her adoptive families have is obvious.

They not only accept her but support her and her additional needs. She is unlikely to be of practical benefit to them, but evidence suggests that orcas are hard-wired to care for the group by looking after all individuals, regardless of whether they are a burden.

Research in 2007 found that, in an orca brain, there are specialised 'spindle cells', which are also found in humpback whales, fin whales and sperm whales, but not in smaller cetaceans. These specialised cells are found in the region of the brain that is responsible for speech, social organisation and empathy. It was thought for a long time that humans are the only mammal, indeed animal, to have spindle cells, and that they set us apart as more emotionally intelligent. It appears, though, that orcas not only have as many as three times more, but have had them for longer in evolutionary terms. Their capacity to feel and respond to emotional needs may exceed our own, and their disposition to look after their sick is a beautiful demonstration of their capacity to love.

Stumpy is not the only orca helped by family. Similar stories come out of South Africa about an orca with deformed pectoral and dorsal fins being supported by his family. This young male is unable to hunt his own food but has been witnessed trailing behind his family as they hunted a fifteen-metre-long Bryde's whale, all the while taking chunks back to the young male.

Spike, (or number 49) from the Scottish bottlenose dolphin catalogue, was first seen in 1989. He was frequently seen along the east coast of Scotland for several decades but, in 2019, was found dead, and it only became apparent just how deformed his spine was when he was taken out of the water. In places it had fused together in an S-shape, so swimming must have been incredibly hard work. It was clear from his skeleton that his spine had degraded over a long time-period until it finally broke in two, which was quite likely how he died. Parts had previously broken off and were found floating within his muscle mass. He had clearly not been in a good way, but had persisted and survived, no doubt thanks to the support of his pod members.

Mankind has long believed that we hold an evolutionary edge in showing compassion like this, but we are clearly not alone. It is very feasible that orcas also feel compassion and look after others who may be less able. While it is thought that deformed dorsal fins in nature are a genetic abnormality, it hasn't been confirmed whether it creates any lasting problems for the individual. Therefore, we simply don't know whether our Floppy Fin has to work harder than the rest or whether he is looked after by the others, but he is mostly seen with John Coe and Nicola so perhaps they keep an eye on him. While the likes of Floppy Fin are pretty rare in the wild, collapsed dorsal fins in captivity are all too commonplace.

In captivity, nearly all males have collapsed or collapsing fins, as do many of the females, even though their dorsal fins are much smaller. Constantly swimming in circles, poor muscle mass and an inability to spend much time underwater means that the fins buckle under their own weight.

Videos taken by visitors to SeaWorld when this topic is discussed are easy to find on YouTube. In one, you can hear a SeaWorld employee giving her spiel about dorsal-fin collapse. A lot of weight is laid on the fact that they have the biggest dorsal fins of any marine mammals and are supported only by collagen. You can almost hear her building up to say, 'No wonder they collapse! It is just common sense, so do not be fooled into thinking it is because they are held in tiny tanks.'

She goes on to say scientists think the condition could be genetic and, since this male in question has fathered fifteen calves for SeaWorld, it is no wonder that they all have the same fin. It is a case of simple inheritance.

The choice of practised wording is careful not to explicitly state that they are *not* responsible, but lead the paying guest to believe that collapsed dorsal fins are bound to occur. That it is no big deal and really not a result of the conditions the whales are kept in.

Extract from YouTube clip of SeaWorld employee:

> dorsal fins are all unique . . . some have flopped over dorsal fins and some have perfectly straight dorsal fins, so basically I am finding that this misconception that people might have is that they have some control over their dorsal fins or that it is as a result of where they live and it absolutely has nothing to do with that whatsoever.

The clip goes on to say that orcas are social animals and often fight to determine social hierarchy, that sometimes this can lead to injuries such as dorsal-fin collapse. This could not be further from the truth. Yes, orcas are highly social, but their group structure is not determined by fighting. There is an adult female, the matriarch, who is the boss of the group surrounded by her offspring, her offspring's offspring, and her siblings. Sometimes she can have three generations of her own bloodline living together all day, every day for her entire life in a tight-knit cohesive group that work together to hunt for food, and share it out. Mothers have been shown to share nearly 90 per cent of each fish they catch with their family, never expecting anything in return, and she will continue to do this for life. Their days are spent hunting, feeding, resting, playing and sleeping together. Rarely are there incidences of territorial fights within a family.

Dorsal-fin collapse as a result of fighting has never been documented in the wild, although fighting is commonplace in captivity. Research from the wild paints an overall picture of peaceful encounters and co-operation and comes from over thirty years of intensive research, including hundreds of thousands of hours of observation.

The skin of an orca is incredibly sensitive, which is why they spend so much of their time touching each other and maintaining physical contact. They clearly enjoy their sense of touch. However, this sensitive skin also means it can get damaged easily, and many have scars from rake marks, cuts from jagged rocks or from propellers.

Experts at the Marine Mammal Research Centre in Vancouver told me they didn't think the rake marks were a form of social bonding, play-fighting if you will. They are more likely to be the result of inter-group mixing and scuffs. However, the cuts are usually fairly minor, leaving faint scars that fade or disappear over time. It doesn't appear to be the case that one orca is trying to physically harm the other or do lasting damage, more of a warning bite between individuals from different families. It is not thought that quarrels are commonplace within a family matriline.

Sadly, the same is not true of orcas in captivity, with aggression between whales held in the same tank commonplace and sometimes brutal. As can be aggression between orca and trainer.

The film *Blackfish* showed the aggressive tendencies of captive orcas, in particular a male called Tilikum, who was responsible for the death of three people. Tilikum was also SeaWorld's biggest breeder whose sperm was used to fertilise many of their females, resulting in twenty-one next-generation orca who will never know what it feels like to swim in the open seas.

One of the many problems with captivity and breeding is that orcas are not kept in their matriarchal groups. In the wild, post-menopausal females are instrumental in teaching their children how to raise their young and help with bringing up and educating their grandchildren. In captivity this doesn't happen. All too often, offspring are removed from their mother's side. Also, mothers are forced to breed far too young, being only teenagers themselves. They have not been taught how to parent.

The young are removed, shipped to other parks and put with other whales that might have been taken from another part of the world. They are not able to communicate with each other, so the young don't stand a chance when it comes to knowing how to behave. Former trainers describe them as permanent children throwing tantrums, but with the strength of an adult. A deadly combination.

I know from my own two children, one just grown out of the toddler stage at the time of writing, and another just entering, that they can certainly throw their weight around when things don't go their way, but they have their father and me to teach them how to behave. Hopefully, they will grow up to be patient and have control over their emotions.

Orcas in captivity are deprived of similar species education and social guidance. Tilikum did not receive this guidance and endured a very long life in a concrete tank. He was captured in 1983 in Iceland, and therefore could have belonged to the population that frequent Shetland and the Orkney Islands in the summer months. He was just two years old when he was captured and moved to Sealand of the Pacific (British Columbia). At night, he and the two females he was with were locked in a tiny metal tank in total darkness, and there was a lot of aggression. The two females were dominant, attacking him in the darkness and raking his skin.

Since he was understandably unwilling to go into the night-time holding tanks, the trainers withheld food through the day and only fed him at night, so that he would be encouraged to enter. He must have been living a life of perpetual anxiety and fear, and it was there, in 1991, that he killed 21-year-old Keltie Byrne. In 1999 he killed again: Daniel Dukes, who had stayed behind after closing hours. No one really knows how it happened but Daniel lost his life. In 2010 Tilikum brutally killed his trainer Dawn Brancheau at SeaWorld Orlando, and it was this death that led to a public enquiry, caused the film *Blackfish* to be made and inspired the subsequent public disgust at how orcas are treated.

Tilikum died in January 2017, after thirty-three years of living in a concrete tank. For the last few years, after the death of Dawn, he was kept mostly in isolation for fear of a repeat attack. Trainers are no longer allowed in the water with orcas, and now have to stand behind barriers to prevent being dragged in. This is absolutely the right decision for trainer safety, but I find it sad that what was possibly the only enjoyable part of the orca's day, the contact they had with their trainer, has now been removed.

I don't believe that Tilikum was naturally aggressive. He spent most of his time working co-operatively with humans, but did he go to the 'dark side' in rare moments of madness and frustration? Yes, I think most certainly. Most people like to think of the orca and their trainers as friends, but who is to say they don't view them as prison guards and, if they see an opportunity for revenge and payback, they take it.

Two scenes in the film make me cry every time I watch it. The first is when they describe how capture events in the 1970s took place. First, the corralling of mothers and babies before separating them. The mums were free to swim away because, being too large and heavy to transport, they were not desirable to the trade. However, they did not leave, and the video shows them remaining close, vocalising and trying to get to their babies.

The little ones were then taken away from their family and put to a lifetime sentence in a concrete tank. It is absolutely heart-breaking to watch that scene. One of the divers from those captures, back when we knew very little about orcas' incredible emotional intelligence, said that as he heard the females vocalising he could tell it was in desperation and grief. That what he was doing was morally wrong.

The other scene is when former SeaWorld trainers describe the day that Kalina was taken from her mother, Katina, in SeaWorld. Kalina was just four years old and had spent every moment of her life with her mother, just as she would have in the wild. However, it was felt that she was becoming 'disruptive' in the shows and so she was moved.

As soon as she was taken from the water, Katina started vocalising in a way that had never been heard before, from a corner of the tank where she lay motionless. When her vocalisations were played back to a scientist, they were determined to be long-range sounds at a frequency that would travel far and wide in water. She was calling to her daughter, desperately hoping that she would call back. The former trainers all struggled to re-tell that story, their pain all too visible.

Katina's grief was obvious, and I think some of the trainers lost faith in SeaWorld at that moment. They loved these animals and seeing her grieve like that was just too much. As a mother myself now, it affects me even more to watch that scene and to think about what she must have gone through that day, through the coming days and probably every day since. I could not imagine losing my children, the very thought makes my chest tighten in panic and pain, and I strongly believe that orcas have the same, if not more, depth to their emotions. We are not alone or special in our ability to feel, and knowing that this was done repeatedly, to satisfy human curiosity and make a fortune for the owners, fills me with despair.

SeaWorld and companies like it, *have* played an important part in orca conservation. Initially, they captured our hearts when it comes to orcas. Once a feared and even detested animal, shot at and hunted, the captive performers opened our eyes to their nature and we began to see them in a new light. These huge creatures, some over 8,000 pounds, were willing to work alongside humans, being gentle, co-operative and caring. Within a generation, we had done a 180-degree turn and fallen in love with these gentle giants.

As Sir David Attenborough once said, 'No one will protect what they don't care about; and no one will care about what they have never experienced.'

Changing public perception has been a crucial step in the research and conservation of this species. However, this has come at an incredibly high cost to the welfare of the individuals captured, individuals lost in the captures and their families left behind.

Speaking of research, SeaWorld is also one of the biggest funders of wild orca scientific studies and also a huge funder for rescue and recovery efforts worldwide. They also financially support rehabilitation centres for many marine mammals and, at the point of ending their breeding programme, invested a huge amount of money into rescue and rehabilitation. That being said, now really is the time to draw a line under this industry of animals in captivity for entertainment and free the remaining cataceans, even if only into an open-water coastal sanctuary site. There is no conservation merit in studying such confined animals. A concrete tank, however large, does not in any way replicate their natural environment. Behavioural studies are therefore inappropriate.

The Western world has at last woken to the abhorrent practice of keeping orcas in captivity, whether by capturing them from the wild or through intensive and invasive artificial insemination programmes. Thanks to the film *Blackfish*, and new regulations to prevent contact between trainer and orca, SeaWorld has had to scale back its numbers and it is now illegal in much of the world to capture or breed orcas. SeaWorld now relies solely on existing stock, under increasing pressure for their release.

The same cannot be said in China where the captive 'entertainment' industry is gaining momentum. China is now rapidly acquiring more orcas from the wild, mostly from Russian waters, to satisfy an appetite of marine-mammal entertainment.

In 2018, almost a hundred belugas and orcas were captured near Vladivostok, Russia and held in what can only be described as prison for a year. After much public outrage and many protests, they were released back into the wild, some in poor condition and far from where they were caught. It is not known what the long-term outcome for these animals will be, but better than a life in concrete confinement.

China currently has over eighty marine theme parks, with more under construction. Over a thousand cetaceans are held in these parks. Globally there are over three thousand in marine parks. That means over three thousand highly intelligent sentient beings living their lives out in confinement, boredom and, in many cases, unsafe environments. Separated from family, they have lost their identity, their community, their culture and their purpose. This is most acutely felt by the orca, whose social hierarchy is so complex, and emotional development so impressive, that the sense of loss must be almost too much to bear.

It must stop. Enough is enough. I urge everyone to think twice before going to see marine mammals in captivity, especially such highly developed creatures as orcas. Go whale-watching instead, with a reputable company and see them in the wild, where they should be. I promise you that you will have a more enjoyable and rewarding day. You will still be enthralled, amazed and entertained. Seeing these enigmatic creatures in their home waters, milling around, playing with each other or with a piece of kelp, breeching for the sheer fun of it and tail lobbing just because they want to, is truly something to behold.

Even better than this, visit one of the sites listed in The Hebridean Whale Trail in Scotland (see the chapter 'Looking Forward') to discover great land sites where you can chance your luck at seeing whales and dolphins from the land with zero impact on the animals.

PUFFIN

Hebridean hubbub

PUFFIN (009) IS A FEMALE ORCA who was first catalogued by HWDT in 2000. She has only been sighted a couple of times since then, but it is thought that she used to mostly hang out with Comet, one of the males I was lucky enough to encounter in 2014.

When people think of puffins and the Hebrides, they are likely not thinking of the orca from the West Coast Community, but the other charismatic black-and-white creature going by the name Puffin. I had hoped that we might get to see these comical-looking birds with their stout little bodies and oversized heads. Being part of the auk family, they are related to penguins and, in much the same way as penguins, they seem rather clumsy on land, but once they dive into the ocean to catch their lunch of sandeels and herring, they are swift and elegant. I always thought they were big birds, but it wasn't until this trip that I learnt from Tim, our skipper, that they are a mere eighteen centimetres tall when fully grown. What they lack in size they make up for in character and, for me, are a real standout bird of Scotland.

We were incredibly lucky during our trip that we had a window of good weather to head out to St Kilda, some sixty-odd kilometres from Leverburgh, where we were moored the night before. Not many people ever get the chance to visit these islands, as the journey can be rather perilous and the weather unpredictable.

Tim, having lived his life in the Outer Hebrides and spent most of his time on the water, knew we had a good chance of making it there without too bumpy a ride.

When he said that we could go I was quite excited as this is a great place to see puffins, with over 100,000 breeding pairs there in the summer. Sadly though, for us, by the time we arrived in early September, the puffins had departed back to sea where they spend their winters. They would return the following summer, to the same nest with the same partner, to rear and raise the next round of offspring in their underground burrows and cliff edges.

As for our Puffin, the orca, there really isn't a whole lot known about her. The only story I heard was of a time when she was seen flinging a tiny porpoise repeatedly in the air, possibly to kill it, or perhaps just for fun. Behaviour like this in bygone days earned the orca its old name, killer whale, and its reputation as a voracious predator. In other orca populations that eat porpoises, like the Bigg's orca, it is not unusual to see them chasing their prey to exhaustion and tossing it in the air before killing and consuming it. Quite often juveniles are involved, so there is likely an element of training going on, disguised as play.

When I was an intern at the Vancouver Aquarium, I spent months analysing video and still images taken by drones of many occasions when orcas and porpoises played side by side, interacting with each other and clearly enjoying each other's company, even playing a game of pass the kelp. There were no signs of tension or aggression that I could see, but these were Northern Resident fish-eating orcas who posed no threat. Clearly the porpoises knew they were safe, despite their identical appearance to the Bigg's orcas that also occupied those waters and most certainly hunted them.

There are accounts of the fish-eating Southern Resident

population chasing porpoises and throwing them in the air but, while the porpoises suffer teeth marks from being held in the orcas' mouths, no bites are taken when the Southern Residents are involved. They are not interested in eating them, just playing. The deeper reasons for this behaviour are not known.

This solitary story of Puffin hunting a porpoise is all we have, and even then, the orca's identity is not confirmed. She hasn't been seen in many years and it is thought that she is dead now. Lost before we got to know her, but this is part of the tragedy of the West Coast Community ... or so it seems. Maybe she is still out there. She isn't as distinctive as some of the others, so perhaps she has been seen, even photographed, without a positive identification. We can only hope that this is the case.

The West Coast Community are rarely seen, particularly some individuals like Puffin, which is why so much of their lives remains mysterious. They have no set routes or routines and it is impossible to predict when you might encounter them or where. What we do know from this story and other eye-witness accounts like that of Daniel Brookes with Nicola and Floppy Fin, is that they eat porpoises, and that they have excellent hearing.

Regardless of how noisy the orca may be during the kill, they are silent during the hunt. Porpoises also have excellent hearing and will quickly desert any area when they hear an orca approaching. Orcas have to be tactical hunters, and they have excelled at this, to become the oceans' apex predator.

The tactics Puffin and her family use to catch their dinner likely change depending on what they are hunting. Based on what we know from the Bigg's orcas of the Pacific Northwest, a seal is a quick kill, and while they might circle around seal haul-out sites, usually rocky outcrops, the hunt and kill usually occur in open water where the seals cannot escape by hauling out onto land.

Harbour porpoises are bigger and orcas tend to hunt them in small groups usually of three or four. Working as a group, they chase the porpoise until it is exhausted, which is likely to involve a few high-speed chases, and the hunt often ends with the porpoise being thrown into the air repeatedly, as was seen with Puffin. It's not really known why they do this. Perhaps to stun it, or break its back so it can't swim away. Maybe it's just a game, a victory dance of sorts. Whatever the reason, it is a long, slow, tortuous death for the porpoise, but death in the wild is rarely a quick and painless affair.

We know from Puffin, Nicola and Floppy Fin that the orcas of the West Coast Community eat porpoises. We also know that they occasionally hunt larger prey, such as minke whales. Again, based on our knowledge from the Bigg's orcas in British Columbia, we can speculate that they work in groups. Taking down a large whale takes coordination, communication and sophistication, in which their acute hearing comes into play, and this is where they get their names 'the wolves of the sea'.

It can take hours and many miles of chasing and submerging the minke to finally exhaust it. While orcas can swim faster, up to 45km/hr compared to 30km/hr for the minke, they can't sustain these speeds and therefore take it in turns, continually swapping the lead. They will force the minke into inlets if they can where, once trapped, the orcas can repeatedly ram it, or submerge its head and blow hole until it drowns. Having worked as a team, they then dine as a team, sharing the food. Regardless of the prey species, orcas, both resident and Bigg's, make sure everyone in the family eats, including those too weak or too young to have taken part in the kill.

This is a beautiful example of their propensity to support and include all members of their kin, and a telling demonstration of their incredible social connections. Everything they do is for the greater good of the family although, as I said, death in nature is hard to stomach. Not easy to write about, or read about, it is even harder to witness, but it is all part of the circle of life. It is hunts like this that have given orcas bad press and the persona and reputation of feared 'killers'. I hope we can all see that, while it is tough, it is nature and that, rather than fear the orca, we should view its highly co-ordinated, complex and social hunting techniques as measures of intelligence.

Bigg's orcas, and the orcas of the West Coast Community, are known to be pretty vocal after a kill, but are silent from the start of the hunt through the approach, utilising their incredible hearing to detect their prey. Porpoises are very quiet, surfacing for a breath in near silence, barely noticeable apart from a quiet 'pfft' as they expel their used air, creating a small ripple in the water as they do. Clear your throat and you miss it. To detect them, Puffin and her family remain equally silent, but constantly alert and listening through any noise the ocean may support ... and the ocean is never silent.

You could be forgiven for thinking that the Hebrides is a quiet place. It certainly seemed so when I sat on the deck of the *Silurian* first thing in the morning, coffee in hand, mist still clinging to the calm ocean. We anchored each night in quiet bays, protected from any harsh winds, and the stillness in the mornings was so peaceful. Birdsong, a gentle breeze and the lapping of the waves on the side of the boat came at me like whispers. I loved those moments before the rest of the crew woke up and we started our day.

In moments like that, it is easy to forget that the ocean beneath the surface resounds with marine mammals vocalising, schools of fish darting here and there, crashing waves and the clicking of snapping shrimp. Throw in the sounds of cargo ships, cruise ships, fishing-boat engines, chains and anchors, fishing gear, motorboats and jet skis, the cacophony of noise grows louder every year and makes it harder for groups of marine mammals to stay acoustically connected.

In Nicola's chapter, we read how recent studies have likened the impact of large vessels to intercity trains and how orcas cease communicating when a large ship passes. It is simply too noisy to continue their conversations. We have created an environment so loud it is bound to have dire impacts on their family bonds and sense of cultural identity. This increase is, sadly, a general problem that affects most of the oceans, particularly areas around the coast-lines. Where there are humans, there is underwater noise. However, in the Hebrides, there are noise issues that pose more risk than mere shipping noise and fishing boats.

Although there are over 30,000 species of fish in our oceans, in the UK we tend to eat just a few, mostly cod, haddock and salmon. The North Sea cod population was pretty much depleted before strict fishing limits were set, but now cod stocks have recovered a little, although not to the healthy levels of yesteryear. Our love for salmon is a significant but different problem environmentally, with not nearly enough wild salmon available to satisfy our enormous appetite. In answer, salmon farms have been established around the coast, providing fish for sale here and to countless countries abroad.

Scotland is home to some two hundred such farms, nearly all on the west coast, in scenic, quiet, calm bays and inlets, with majestic mountains as a backdrop. All you can really see of them is their distinctive round pens covered in netting with a large boat tied next to them. Some will say they are a blight on the landscape, and I agree the scene would be more attractive without them, but they don't really detract from it all that much. Unfortunately, they are not quite as innocent as they seem.

Aquaculture is a huge industry generating revenue for Scotland of around £2 billion per annum, and creating jobs and incomes for eight thousand people, many of whom live in the remote north and west of the country, where there are few other jobs. Furthermore, fishing is a way of life and part of the culture of the Hebrides. So, as you can imagine, there is a lot of motivation for the fish-farm owners to protect their incredibly lucrative industry.

Set up to satisfy human appetites, these farms also attract other hungry visitors. Seals absolutely love salmon and are a real pest to the farmers, and Acoustic Deterrent Devices (ADDs) are now deployed in more than 140 fish farms across Scotland with the aim of scaring them off.

ADDs were developed some decades ago with the first models using underwater speakers to play recordings of orcas. Hearing their main predator, seals would quickly leave, but this didn't last long. The seals cottoned on. So, a particularly unpleasant sound was developed and played continuously underwater, successfully deterring the seals. However, the pings emitted from the ADDs proved not just to be unpleasant to hear, but to actually be harmful to wildlife, particularly marine mammals with their sensitive hearing. Whales, dolphins and porpoises experienced serious pain and lasting damage when they came too close.

Marine mammals exposed to loud underwater noises are known to exhibit reduced hearing sensitivity, referred to as a 'threshold shift' (TS), which can either be temporary (TTS) or permanent (PTS), and can experience changes to breathing, nursing, breeding, resting and mating, leading to increased energy consumption and increased vulnerability. The use of ADDs around Scottish fish farms is set to increase as in 2020 a Bill was passed by the Scottish Parliament that made it illegal to kill seals, bringing us into line with the American legislation and paving the way to the export of vast quantities of Scottish salmon.

If the farmers are banned by law from killing the pesky seals, they will be even more motivated to keep them away in the first place. Research has shown that ADDs are already creating significant, chronic and widespread noise pollution along the west coast of Scotland with no likelihood of any reduction. No licence is required.

When I was conducting marine mammal-surveys with the Hebridean Whale and Dolphin Trust, we always had the hydrophone in the water, trailing behind the boat with one person stationed in the saloon with headphones on. It was a relaxing job most of the time, listening to the snapping shrimp and the clicks and whistles from playful passing dolphins. Boat sounds weren't so welcome, but nothing sounded worse, indeed actively hostile, than when we passed a fish farm with its incessant pinging. Orcas will certainly hear them and potentially be harmed by them. They do not have an appetite for salmon, so they don't pose a risk to the fish farmers but, of course, we know that they eat the sort of marine mammal that might gather there. So, ADDs could impact our orcas in two ways: by causing them pain if they get too near, and by interfering with their hunting. Research has shown that ADDs not only force orcas to abandon an area but also displace them from feeding areas more generally.

The more we deplete our natural resources and turn to artificial food generation, such as fish farms, the more we displace our marine life from their natural habitat range, and the cumulative impacts of ADDs at multiple fish-farm sites in the Hebrides generates adverse conditions across vast areas. First, we remove their food source, then we take away their homes. It's a depressing scenario and, surely, we can find better ways to live side by side more sustainably.

Fish farms have been highly criticised for their negative impacts on the environment. To counter the risks of overcrowding and disease, chemicals such as hydrogen peroxide are pumped into the cages to help prevent infections and sea lice. These chemicals persist in the water column and build up in food chains, reaching invertebrates on the seafloor and affecting many species in addition to the salmon. As we saw with Lulu, chemical pollution can have incredibly damaging sub-lethal effects and it is likely we will see its legacy last for decades and longer.

Then there is the food they are given, which is not wholly comprised of fish such as the salmon would naturally eat, but supplemented with vegetable and palm oils to as much as 50 per cent. The highly concentrated waste that the salmon produce leads to excessive nitrogen levels around the pens, causing eutrophication and toxic algal blooms, with the areas beneath and around the cages turned into waste zones where natural wildlife struggles to exist.

There is a lot of opposition to fish farms, but while demand continues to generate employment and revenue it is unlikely to change. Working with the aquaculture industry to help minimise impacts is therefore of supreme importance.

Salmon farms are also big business in British Columbia where ADDs have been banned since 2000. Instead, they use 'anti-predator nets' which are simply strong, weighted nets that surround the cages to keep the seals out and the salmon in. It's a simple but effective practice that eliminates the need for harmful ADDs. With the huge profits that salmon farming brings in, it must be possible to invest in these nets and remove this source of pollution for the greater good of all cetaceans.

Scottish Environment Link is a forum of voluntary environmental committees supported by organisations such as Hebridean

Whale and Dolphin Trust, Marine Conservation Society, National Trust for Scotland, Scottish Wildlife Trust and Whale and Dolphin Conservation. They released this statement on the use of ADDs in Scottish waters in 2020 as follows:

> Scottish Environment LINK considers the use of Acoustic Deterrent Devices (ADDs) by the aquaculture industry in Scottish waters poses a significant and unnecessary risk to cetaceans. LINK believes regulation and management of ADD use by the salmon farming industry, with the ultimate aim of phasing out their use entirely, is required to ensure protection for cetaceans and full compliance with European Protected Species legislation.

It is great to see pressure put on the aquaculture industry to phase out the use of ADDs. What a different underwater soundscape there will be without their incessant pinging.

Every spring and autumn, the relatively peaceful waters of the Hebrides become the stage for quite a show. The Joint Warrior Exercises carried out by the Royal Navy are practice for dealing with military conflicts that might occur in the future and part of a collaborative effort involving the Royal Air Force and the Army. Although led by British forces, it is part of a huge multinational training programme involving thirteen other nations: Belgium, Canada, Denmark, France, Germany, Latvia, the Netherlands, Norway, Spain, Turkey, Japan, UAE and the US who, over the course of two weeks, practise and prepare for crises and conflicts such as territorial disputes, piracy and terrorist activities. There is no denying that we need these defences, and for them to be effective this training is necessary, but they come at a cost to the orca and other marine mammals.

The greatest concern for the West Coast Community is the use
of military sonar. Sonar stands for 'sound navigation and range'. It
was designed, at least partly, to detect submarines, and can be either
passive or active. In active sonar, a pulse is sent into the water from
a surface ship or from a submarine, and any solid objects will
bounce the sound waves back. This provides a picture of the size
and location of the target object; pretty handy if you are looking
for a hostile submarine that can't be seen deep under water.

Orcas have their own, biological sonar, otherwise known as
echolocation. Like other dolphins, they emit pulses from the
melon in their head and any reflected pulse enters the jaw into a
fat-filled cavity. From that information, an image of the object is
formed in the orca's brain. Bigg's orcas, and therefore likely Puffin
and her family, echolocate infrequently, only enough to maintain
contact and communication. Not enough to alarm their unsus-
pecting prey.

The concept is the same though. They send out pulses that are
bounced back, and from them a picture is created of the underwater world. However, the problem with military sonar is that the
pulses emitted are incredibly loud and travel a long distance.

All marine mammals, not just orcas, have sensitive hearing and
loud noises can cause them to change their behaviour. They can
be injured by the noise, which can even result in death. Noise
pollution can result in damage to auditory tissues, lesions in the
jaws and inner ear and cause nitrogen bubbles to form in the
blood, giving marine mammals the 'bends'.

The bends, or decompression sickness, is something we typically associate with deep-sea divers and, for a long time, it was
thought that marine mammals didn't suffer from it as they make
repeated dives without ill effect. However, it appears that this isn't
the case. When they are at depth, nitrogen gas builds in their
bloodstream and tissues. When they surface slowly, the gas has

time to move back to the lungs and can then be expelled in their breath at the surface.

Come up too quickly, as they might have to do to quickly evade sonar and other unpleasant sounds, the nitrogen doesn't have enough time to diffuse back into the lungs. With each metre they rise, the pressure reduces and the nitrogen-gas bubbles in the blood and tissues expand and cause great physical pain.

Military sonar, along with seismic surveys and underwater explosions, is acute noise, louder and more damaging than the chronic sub-lethal noises of general vessel traffic or ADDs. This acute noise can lead to permanent hearing loss which means, for orcas, that they may not be able to find food, reach breeding and feeding grounds, locate mates and other individuals within their group, or maintain their balance.

Military sonar is, at best, disruptive to marine mammals, causing them to avoid certain areas and abandon feeding, but it can also be lethal. Research has linked the military testing of sonar to mass strandings, particularly of deep-water species such as Cuvier's beaked whales, which ascend too quickly and suffer decompression bends from which disorientation can follow. Whole family groups can strand on shore and die. The evidence has been mounting and the links are now clear, with scientists calling for a complete ban.

The Royal Navy is required by law to undertake risk assess-ments, carry out mitigation studies and do all they can to mini-mise any impacts. Before each Joint Warrior exercise, officers must submit an Impact Statement in compliance with the Wildlife and Countryside Act 1981, Nature Conservation (Scotland) Act 2004, The Offshore Marine Conservation (Natural Habitats) Regulations 2007, Marine and Coastal Access Act 2009, The Conservation of Habitats and Species Regulations 2010, The Marine (Scotland) Act 2010 and the Marine Act (Northern Ireland) 2013. A huge number of policies and practices are in place to safeguard the environment and, before sonar is engaged, a thorough Sonar Risk Assessment is conducted. Marine Mammal Observers (MMO) seek to detect any marine mammals present in the area, and all sonar use is kept to daylight hours so there is visibility to detect marine mammals. Overall, it is deemed by the Royal Navy that the environmental impact is 'Minor'.

Is this true and are these mitigation methods adequate? This is obviously of great concern in the Hebrides where there is an abundance of marine mammals, and the fact that military testing is conducted in Special Areas of Conservation for porpoises and known cetacean hotspots leaves me incredulous.

In areas supposedly dedicated to the protection of our marine mammals, the timings of Joint Warrior were altered to coincide with the times of the year when cetacean sightings are high. The HWDT use the *Silurian* to monitor the effects on marine life and have been doing so since 2009, including through the COVID–19 pandemic in 2020, but that amounts to one 61ft ketch against fifty-eight aircraft, sixteen warships and some 3,700 military personnel. In reality? As onlookers it is impossible to know what is really going on.

With Lulu, we read about the legacy of PCBs, a form of chemical pollution that hasn't been in use for forty years but persists in damaging concentrations in our oceans. We must take action on them, but there is no quick fix. In comparison, noise pollution we can do something about. If we turn off the noise, the pollution goes away. If we can get the use of ADDs banned and stop the use of sonar testing, we can remove two of the biggest noise polluters in the Hebrides.

In 2020, the year many of us would like to forget, the year the world stood still, we took refuge in our own homes as COVID–19 swept across the globe, and nature got a chance to breathe, to pretend the world is as it was before we came crashing in. Cruise ships remained at anchor, recreational vessels stayed in their docks and moorings and the ocean noise was dialled down. Many studies will no doubt show how nature adjusted and thrived. They will show that we need to tread more quietly on our planet and especially our underwater world, and leave the song of the ocean to be sung by nature.

AQUARIUS

Folklore and fables

AQUARIUS (008) HOLDS A VERY SPECIAL place in my heart. The first orca I ever saw in the wild, he is still going strong today with John Coe, and every reported sighting of him fills me with joy. He is an adult male, first catalogued in 2004, several years after John Coe and Nicola were added to the HWDT catalogue. However, this doesn't mean he is either new to the group or younger than the others as he was already a fully fledged adult with a huge, sprouted dorsal fin. He was at least in his teen years.

He simply hadn't been recorded before 2004, but has been seen many times since, and is one of the only two remaining members we can be sure of as only he and John Coe have been seen since 2016. The pair caused quite the media frenzy in 2021 with sightings off Cornwall, in the English Channel, on the east coast of Scotland and, of course, several sightings in the Hebrides. Each sighting of them, likely the last remaining orcas from this group, is so very precious. We all hope that we have many more sightings in the years to come.

Aquarius was one of the two males I saw early on in my trip in 2014 as we passed Neist Point Lighthouse. I had joined the *Silurian* research vessel in the hopes of seeing the West Coast Community, but didn't really believe it would happen as, with just a few left and a whole lot of sea to explore, the chances of randomly bumping into them is very slim. But we did, and it was every bit as magical as I had imagined. Even more special was that I was on lookout and the first to spot him. The adrenaline rush as the two orcas approached, Aquarius and Comet, their huge muscular bodies slicing the water with ease and elegance, was intense. Seeing them was the inspiration for telling their story and writing this book.

Given their powerful presence, their intelligence and the intrigue that surrounds them, it is no wonder that orcas have been subjects of cultural significance, present in stories and folklore for thousands of years.

Orcas have had a bad reputation with fishermen and seafarers for a long, long time. In the first century AD, Pliny the Elder wrote: 'a killer whale cannot be properly depicted or described except as an enormous mass of flesh armed with savage teeth'. In 1874, the whaler Captain Charles Scammon wrote: 'in whatever quarter of the world [killer whales] are found, they seem always intent on seeking something to destroy or devour'. In 1973 the US Navy diving manuals described the orca as: 'extremely ferocious', warning that 'it will attack human beings at every opportunity'.

Their scientific name, *Orcinus orca*, demonstrates that they have always been linked to fear by man, for Orcus was a Roman spirit of the underworld. Their high intellect means that they have learnt to interfere with fishermen and their activities, taking fish off their lines, getting in the way of whalers and generally making a nuisance of themselves. Their escapades have been the stuff of legends all around the world, wherever they are found. Including Scotland.

The Hebrides have a long and rich human history which has created a unique culture, with evidence of human habitation dating back eight thousand years, when it appears that the forests then covering the islands were burnt to make room for deer, forever changing the landscape. The Callanish stones, stone circles made from raw gneiss rock, were built some five thousand years ago. To put that in context, Stonehenge, one of the most iconic historic monuments in the world, was only constructed around 2,500 years ago. The Scots arrived from Ireland in the years after Christ, bringing with them the Gaelic language. The Celtic monk Saint Ninian arrived in the fifth century with Saint Columba following in the sixth, bringing Christianity with them. The Vikings arrived in the ninth and the Hebrides fell under Norse control. Around 1266, they were returned to Scotland following the Treaty of Perth, which was then followed by centuries of infighting between different clans. The disaster of Culloden meant the end of the clan system and the Clearances followed, a time when sheep were given more importance than humans and entire villages from the islands and highlands were evicted to the coast and to emigration.

I am no historian, and would not wish to oversimplify the melting pot of cultures and events that have shaped the lives of the people of the Hebrides, but stories from those early times have been passed down the generations. Folklore is an integral part of Scottish culture.

Many people think of of a ceilidh (pronounced 'kaylee') as being simply a fun dance, much like the barn dancing I did as a child, but a ceilidh is more than just a dance. Originally, it didn't really have anything to do with dancing but was any form of social gathering where songs were sung, literary matters discussed, poetry and stories recited that told tales of the land, the people and their history. Ceilidhs have been a way to pass down culture and heritage in an informal and social way from one generation to the next.

One of Scotland's national animals is the unicorn, a mythical creature loved by little girls the world over, my daughter included. Known in Scottish Gaelic as *Biasd na Srogaig*, the unicorn is a horse-like creature with a horn at the front of its head but, while it might look like a horse, it also bears resemblance to the marine narwhal, which has a huge tusk, a modified tooth, at the front of its head and is often referred to as the unicorn of the sea. However, narwals are not a species that is seen in Scotland.

Why the unicorn? No one really knows. However, it is a creature associated with purity, innocence, power and masculinity and, while not real, it symbolises the Scottish love of folklore and the mythical.

Given that the Hebridean Islands are surrounded by the sea, it is no surprise that marine-based tales and myths are so common. The highlands and islands are rich with stories involving mermaids, kelpies, sea monsters and shapeshifters, the most famous being, of course, the Loch Ness Monster. The serpentine creature that lives in the murky waters of Loch Ness is linked to the Hebrides through Saint Columba.

Loch Ness is large, deep and long. The largest body of water by volume in the British Isles, it contains more fresh water than all the lakes in England and Wales combined. Nessie (or *Niseag* in Scottish Gaelic), as the monster is affectionately known has been discussed and disputed for many years. There are many anecdotal stories about her but, due to a lack of credible photos, she is thought by the scientific community to be no more than a myth. She was first reported back in AD 565 when Saint Columba, the most famous settler on the Isle of Iona, visited the east coast of Scotland via the River Ness. Importantly though, it was in the river and not the loch where he spotted a large water monster heading towards a swimming man. Saint Columba ordered the water monster to leave the man alone and the man's life was saved.

A logical explanation could lie in the fact that the River Ness connects with Loch Ness at one end and at the other end goes out to the sea via the Beauly Firth. As we saw in Comet's chapter, large marine mammals such as humpback whales (*Megaptera novaeangliae*) and minke whales (*Balaenoptera acutorostrata*) do enter rivers at times.

Other people in this region would have been unlikely to recognise a marine mammal like that, but Saint Columba, from the Isle of Iona, might well have encountered minke whales as they are frequent visitors to those waters. Adding merit to the idea that this sea monster was a whale is that, at this time, the Latin word for whale, *cetos*, used by scholars such as Saint Columba, was the same word used for sea monster or large creature from the sea. He might have been accurately reporting that it was a whale but was misconstrued by the locals, who would not speak Latin, as saying 'a monster'. Assuming that he had encountered minke whales before, he would have known that they are harmless and did not pose a risk to the man swimming in the river.

You can well imagine that, as this story was repeated, details would be altered in the telling until this was a sea monster in Loch Ness. So, it is quite likely that Nessie has been around since AD 565, but not in the loch itself, and was likely just a minke whale making a brief trip from the Moray Firth into the River Ness. Regardless of this explanation, the countless other stories of the origins of Nessie, the decades of searching with not a single credible photo, believers return year after year in the hope of proving disbelievers wrong.

Nessie isn't the only reported sea serpent in Scotland. There have been reports of large mysterious sea creatures living in lochs in the Hebrides as well, such as 'Searrach Uisge', who has been living in Loch Suainbhal on the Isle of Lewis for over a century and a half, and another creature living in Loch Duvat in Eriskay. Still, Nessie remains the most famous and most searched for.

Saint Columba was involved in another famous tale, which might well involve orcas. The story was told by Adomnán in the seventh century, an abbot at Iona Abbey and a cousin of Saint Columba. Both were originally from Ireland.

Iona sits just to the west of the Isle of Mull in the Inner Hebrides. A tiny island, it is just one and a half miles wide and three long. Many of the stories of medieval Scotland come from this island as it was the centre of Gaelic monasticism for over four centuries and known as the 'Cradle of Christianity' in Scotland. In those times, religious leaders were often the only people to have received formal education and therefore the only ones who could write down their thoughts and musings.

The story goes that there were two monks travelling from the Isle to nearby Tiree. Saint Columba had warned the two holy men that 'monsters of the deep' patrolled these waters and, sure enough, they encountered what would most likely be an orca. Adomnán recorded in his book about the life of Saint Columba 'a whale of extraordinary size, which rose like a mountain above the water, its jaws open to show an array of teeth'.

The description of teeth tells us that they had encountered an odontocete, a toothed whale, but there are many toothed whales that can be encountered in the Hebrides. Harbour porpoises, common dolphins and bottlenose dolphins all have teeth, but were unlikely to have caused much fear and are not whales of extraordinary size.

Sperm whales also have teeth, are exceptionally large and known to visit these waters from time to time. However, the behaviour does not fit as they tend to lift their head just enough to take their blowhole out of the water. It would be incredibly unusual for a sperm whale to rise like a mountain but, I suppose, if you were in a small rowing boat it would look rather large just lifting its head. However, sperm whales have a small lower jaw with teeth, and a huge forehead amounting to almost a third of their entire body, something the monks would surely have mentioned. The teeth would likely be the last thing of note to anyone coming face to face with a sperm whale.

The most likely culprit to have scared the two monks as they travelled between Iona and Tiree was an orca. While a lot smaller than a sperm whale, it is still a creature of extraordinary size, especially if you are in a small boat. Anyone who has seen them in real life or, even better, kayaked with them, will know how small a human is in comparison. Unlike sperm whales, they display quite a lot of surface behaviour, from porpoising and spyhopping to full body breaching. The dorsal fin on a male orca can be two metres high, which would certainly look like a mountain to two men in a rowing boat ... and the description of the teeth fits.

Orcas have sharp teeth on both the upper and lower jaw, numbering some forty-eight pairs in total, which would definitely have caught the monks' attention. The only thing that is strange about this story is that its colouring goes unmentioned. Orcas are incredibly distinctive with their black-and-white bodies, but maybe the monks were too distracted by the teeth and sheer size to especially note the colours.

What is incredibly interesting about this story is that, although it was first recorded in the late seventeenth century, it would mean that orcas were present in the Hebrides over 1,400 years ago. The HWDT began recording in the 1990s, and we know from Comet that the West Coast Community has been around at least since the 1950s. Before then, it wasn't thought that orcas were much present in Scotland. Lari Don, a Scottish folklore storyteller, believes from her research that orcas do not feature much in Scottish folklore as they didn't come into coastal· waters until after the relatively recent herring crash.

Herring-eating orcas are now pretty common on the northern isles of Orkney and Shetland, but they are not frequent visitors to the Hebrides. The West Coast Community members are not herring eaters at all, shunning fish that other marine mammals eat, instead going for bigger prey such as porpoises and even minke whales.

They might well be referred to as the 'great beast of the ocean', or the *cionarain-crò*. In the mid-nineteenth century, the folklorist Alexander Carmichael took down a story told by 'Old Angus Gunn', which tells of a good and holy man 'in the dark grey dawn of the ages long ago' who came to the Isle of Lewis and settled in the district called Ness. Sadly, the people of Ness were found to be unholy, so he vowed to leave, not caring where he would go to. After prayers to God were offered up, he heard a voice telling him that he must go to the landing place west of the town of Eoropie, in the most northerly part of the Isle of Lewis, where he would find a messenger who would carry him away from this lawless place.

That messenger turned out to be the *cionarain-crò*, a giant sea creature that rose from the depths of the sea to take him away. A rhyme taken down in 1860 gives some insight into the size of this beast:

Seven herrings
Feast for a salmon;
Seven salmon
Feast for a seal;
Seven seals
Feast for a small whale;
Seven small whales
Feast for a Great Whale;
Seven Great Whales
Feast for a cionarain-crò
Feast for the great beast of the ocean.

The story goes that the *cionarain-crò* took its burden safely across the rough winter seas to the uninhabited island of North Rona. With the good and holy man safely on land, the beast disappeared back into the sea. North Rona lies some forty miles north of the

Butt of Lewis and, even to this day, landing is only possible in excellent conditions. Local folklore says that the beast can shapeshift and disguise itself as a small, silver fish when fishermen are nearby. Once in the nets and brought aboard, it changes back and mercilessly devours them. Being a fisherman was and still is a risky job, with many a man losing his life. Perhaps it is unsurprising that, over the years, sea monsters in various guises have been held responsible.

Given that the *cionarain-crò* could change its form from giant beast to a tiny fish, it might also be called a shapeshifter. A kelpie is the Scottish name given to a shapeshifting water spirit, usually one that lives in lochs and pools. Often having a horse appearance, they can shapeshift into human form but, when they do, they keep their hooves. So kelpies are often associated with Satan, and feature in a poem by Robert Burns, the famous Scottish poet. One of the verses in his 'Address to the Deil' written in 1786 goes like this:

When thowes dissove the snawy hoord,
An float the jinglin' icy boord,
Then water-kelpies haunt the foord,
By your direction,
An' 'nighted trav'llers are allur'd
To their destruction.

The kelpies in this poem conjure images of a scary underworld and, indeed, the folklore surrounding mythical mammals of the sea often portray them as dangerous and fearsome.

The Storm Kelpies of the Hebrides, also known as the 'Blue Men of the Minch' are man-like creatures that inhabit the Minch, the water between the northern Outer Hebrides and mainland Scotland. The legend is that they seek out sailors to drown and boats to sink using their powers to create great storms if the sailors are not able to converse in the poetry that they are so fond of. While they have human form, their skin is blue, and they move in the water like porpoises. It is believed by some that their presence in this area is the reason for so many shipwrecks.

Of course, we know that the Minch has incredibly strong tidal currents and the Hebrides are no stranger to adverse weather conditions and storms.

Some legends also portray marine creatures as compassionate and willing to help mankind. Selkies are the Scottish mythical 'seal creatures' that can shed their sealskins, revealing themselves as beautiful women underneath. The selkies fall in love with men who then trick them and steal their sealskins. The maidens long for the sea and their old life under the waves and try to escape back to their marine form – if only they can find their stolen skin.

The most beautiful maiden of the sea though is undoubtedly the mermaid, or the '*maighdean na tuinne*', which has stayed a popular legend in today's culture the world over. Covered in shimmering scales and adorned in shells and pearls, she is shy and gentle and longs for a relationship with mankind.

It is this empathic behaviour that we relate most to and, in many places around the world, orcas in particular are revered and given special prominence. This is most seen in indigenous cultures with a strong and deep connection to the natural world, and an understanding of these amazing creatures that the rest of us are only just waking up to.

The Tlingit are the indigenous peoples of the Pacific Northwest coast of North America, living between the temperate rainforests of southeast Alaska and its border with British Columbia. They have a matrilineal kinship system where a daughter is born into a mother's clan and will inherit her roles and property while living a hunter-gatherer lifestyle. The land and seas in this part of the world offer a bounty of food but none is valued as much as the salmon when eaten fresh in the summer. It is also dried and cured to provide sustenance through the long, cold winters.

The Tlingit regard the orca as one of their most important animals and they have their own name for them, one that most of us have heard of: Blackfish. This was the name given to the hit film *Blackfish* about an orca named Tilikum, who was held in captivity in SeaWorld. The Tlingit believe that Blackfish would never harm a human, which appears to be true as there is no record at least of a *wild* orca harming a human. Rather, it is a custodian of the sea and brings gifts of food and strength to mankind.

A Tlingit legend which has been shared hundreds of times through the generations is the story of Natsilane, which involves Natsilane himself, his evil brothers, sea otters and, of course, the Blackfish.

Legend has it that Natsilane, with his skills, intelligence and general likability, was destined to become chief when he was older. His brothers were insanely jealous and they tried to kill him by taking him out to sea and throwing him overboard a great distance from land. In the rough seas, Natsilane began to drown but sea otters came to his rescue, floating him to a nearby island, the distance back to mainland being just too great for the small otters. They continued to care for him by showing him all the island's best hunting grounds and providing him with seeds to plant. These seeds grew into the huge trees that are now native to the Pacific Northwest, such as firs, maples, larch and cedar.

Using one of the yellow cedar trees, Natsilane carved a totem depicting a huge fish, which he left on the shore for the otters to find. When he returned the next morning he found that the fish was gone, replaced just offshore by Blackfish, the first orca. The otters had turned the wooden fish into Blackfish to help him.

Natsilane then boarded a boat of his own making and set off with Blackfish as his guide. Approaching his old home, he saw the brothers who, years before, had tried to kill him, and no doubt thought they had succeeded. He ordered Blackfish to destroy their boat and kill them in retaliation and, with that done, ordered Blackfish not only to never harm a human being again, but also to always help a human in distress at sea.

The story is told to pay respect to the orca and show what can be achieved when humans and nature work together in harmony.

The indigenous people of Canada, south of the Arctic Circle, are collectively known as the First Nations, which are then separated into distinct tribal regions, such as the Kwakwaka'wakw, and Nuu-chah-nulth, who are both on Vancouver Island, and further divided into tribal groups such as the Tla-o-qui-aht, the Kwakiutl and the Haxwa'mis. The Kwakwaka'wakw refer to the orca as *maxinuxw*, which means 'side by side tribe' as they travel in groups. The Nuu-chah-nulth, who live on the west coast of Vancouver Island, around Tofino and Clayoquot Sound, have their own name for the orca, *kakawin*. Others give them the common name 'black fish'.

As we read in John Coe's chapter, there are three main groups of orca in the Pacific Northwest: the Residents, the Transients (or Bigg's orca) and the Offshores. The return of the resident orca, particularly the Northern Resident orca around the north of Vancouver Island, is warmly welcomed by the Kwakwaka'wakw people. Their return symbolises the start of the salmon spawning season, a season of plentiful food, but their affection goes beyond the start of the summer feast.

Many First Nation tribes believe that when their chiefs die they shapeshift into other animals, namely the orca, the wolf or the eagle. The Kakawin is the enforcer of the sea, the wolf the enforcer of the land and the eagle the enforcer of the air. This is why many carvings and totem poles bear the images of these three animals.

The Kwakwaka'wakw peoples believe that orcas were the origins of their own tribal groups, living underwater in human form in towns, much as we do above water, but shapeshifting into their orca form when at the surface. Sometimes a male orca would even come onto land in human form to establish himself as the chief of a new tribe.

The origins of the First Nations go back before recorded history as do their stories related to orcas, with evidence of their importance to coastal First Nations found on petroglyphs, which are drawings made on rock faces by etching away at the rock.

In the summer of 2016, when I spent several months living in British Columbia, I visited the site of some historic petroglyphs at Sproat Lake, near Port Alberni on Vancouver Island. A stroll from the car park takes you to the edge of the lake and a floating pontoon on which you can walk to see the petroglyphs from a safe distance. Put another way: it keeps the petroglyphs a safe distance from the tourists. This panel of drawings is called 'K'ak'awin', and is thought to be one of the best examples of a petroglyph in British Columbia, this time depicting a mythical creature, clearly part wolf and part orca with the prominent tall dorsal fin of an adult male. The pictures would originally have had colour, with pigments made from powdered minerals. Ninety per cent of all rock paintings were red.

It is not known how old these particular carvings are, but archaeological remains of the First Nations date back about twelve thousand years. It is unlikely that these drawings are so ancient, but they are thought to be a few thousand years old. It is clear that the orca has held special prominence to the First Nations for a very long time, and that bond shows no sign of waning.

Inuits living in the Arctic have different cultural connections to the orca. In their mythology, Akhlut is a spirit that takes the form of both a wolf and an orca. The origins of the story were not written down and there are several different versions, but the most widely shared starts with a man who was obsessed with the ocean. He spent so much time by the sea that, one night when he returned to his village, they didn't recognise him and banished him. He was so angry and hungry for revenge that he shapeshifted into a wolf. However, his love for the ocean could not be held back so, in his wolf form, he entered the sea and turned into an orca.

This myth has it that he can change his shape to be either fully orca, fully wolf or part of each. When at sea, he lives in peace as an orca but on land as a wolf, he is a vicious and dangerous beast. When Akhlut is hungry, he shapeshifts into the wolf form and comes back on land to hunt, and everything, including the Inuits, are on his menu. The Inuits consider Akhlut to be very dangerous and are wary of getting too close to the edge of the water in case he is tempted to come out of the sea and attack them. It is the tracks of wolves along the beach that scare them most though, and even dogs walking on the beach are considered to be evil. Even though Akhlut is part orca, when in his orca spirit he is not considered dangerous. Again, indigenous communities do not fear orcas, but they maintain a healthy respect.

The Māori tribes of New Zealand have a strong cultural appreciation of whales, or 'tohora' as they are collectively named. Considered to be descendants of Tangaroa, the God of the Oceans,

all whales are revered by the Māori people as guardian spirits sent
to watch over individuals while at sea. They are considered sacred
and to have supernatural powers. Whales will act as guides for
them, and they believe they will lead them to safe places to settle.
As with other indigenous cultures, whales therefore feature heavily in carvings, particularly in the storage houses for their barges.

Whales have had many other uses to the Māori, having been
the source of food and utensils.

In 2014, a very sad event unfolded on a beach in the town of
Tautapere, on the southern end of New Zealand's South Island.
Nine orcas were found beached and dead on the rocky shoreline.
While there are cases of individual orcas suffering this fate, it is
highly unusual for such a large number to beach all together. This
was a huge blow to the relatively small New Zealand population
of orcas, as these nine accounted for 5 per cent of the total population. Even more heartbreaking was the presence of a calf.

This loss was devastating to the conservationists who fight to
protect them, especially Dr Ingrid Visser who has dedicated her
life to them. However, the loss was also devastating for the local
Māori tribes, and a special ceremony was performed by the local
tribe, the Iwi, to bless the animals and wish them well in their
final journey.

Given the huge importance of the orca, permission for anything
other than photographs to be taken has to be given by the Iwi
before scientists can act. Thankfully, they understood the importance of this stranding and the need for answers and gave permission for the scientists to perform a necropsy on the calf. As when
other whales are stranded in New Zealand, the Māori decided the
most appropriate burial ground.

Strandings are, sadly, not a rare thing in New Zealand, which is
considered to be the whale stranding capital of the world. Hori

Parata is renowned as the leading Māori whale expert and has been described as a whale whisperer. Now in his seventies, he has been attending strandings for seven decades and is growing increasingly concerned about the frequency with which they are occurring. He is from the Ngātiwai tribe, who believe that each stranded whale is one of their ancestors returning to their human family, the Māori people. Therefore, the death of each whale is deeply personal and devastating to them all. Each whale is named in a ceremony and prayers are offered up.

Looked on as a gift from their ancestors, they use the bodies, removing bones, blubber, eyes and teeth, for cultural purposes, a right that was legally recognised in 1998. With the increase in strandings in recent years, the Ngātiwai and other tribes believe it is their ancestors' way of letting them know that there is something wrong with our oceans, that we are harming our planet. There is clear frustration from Parata that the indigenous people are not being listened to and that their concern for the whales and what they are telling us are not being heard.

Modern-day science likes to claim all the big discoveries. For example, it was in the 1970s that the team of scientists led by Dr Michael Bigg discovered that individual orcas could be identified by their dorsal fin. They also discovered the Rubbing Beaches in Johnstone Strait where Northern Resident orcas would visit in the summer months. Here, in this remote place, it took years for the scientists to gain enough trust from the orcas to be 'allowed' to witness what was happening. The orcas would rub their bodies against the smooth rocks, perhaps to remove parasites or dead skin or perhaps because it just felt nice, like a hot stone massage.

More than this, these rubbing beaches are a place of importance for social interaction, intimacy and mating. This discovery was a window into their lives – the gentleness between individuals and the importance of physical contact. However, with all due respect to the late Dr Bigg and his team, they were not the first to discover this. The Kwakwaka'wakw First Nations people have known since before recorded history. All around the world, native peoples have a connection to nature that Westeners seem to have lost. Only recently have we begun to realise how dependent we are on our planet and the animals we share it with.

There is so much that can be learnt from the tribes and peoples described in this chapter, and many more besides. Their knowledge and respect for the orca is unsurpassed, and while you don't need to believe all the legends and mythical elements, there is no denying that folklore passed down the generations keeps the allure and intrigue of the orca alive.

OCCASUS

The sun sets in the west

OCCASUS (010) IS A FEMALE ORCA first catalogued by HWDT in 2005 and seen again in 2009 with two of the males, Aquarius and Floppy Fin, as well as two other females, Lulu and Nicola. Moon was assumed dead by this time, which means that these were five of the remaining nine individuals of the West Coast Community, a pretty large gathering by their standards. Usually seen only in pairs or threes, they must have been quite a sight.

Perhaps the reason for this large group was that they were sighted off County Mayo on the west coast of Ireland, a pretty good distance from the Hebrides. They might have been on a hunting trip to seek out new prey, and had they got lucky they might have split the group in two. There were no comments on this encounter from locals, so we don't know if the orca were hunting, and there were no underwater recordings to tell if they were chatting away or staying silent so as not to scare off prey. We know that, at the point of being seen, they were not actively chasing prey or sharing the spoils. They were just five orcas milling around together, checking out a new seascape. What a treat for the lucky Mayoians: to see five orcas together, especially five of this elusive and mysterious community.

Members of the West Coast Community appear to visit Ireland quite frequently. So, it is by no means unusual for them to venture this far from home. Back in 2009, when she was seen for only the second time, Occasus was known only as 010. She was the last of the group to be added to the catalogue, with no new members identified since, and brought the entire recent population to ten. It is suspected that the group was larger in the 1980s before formal monitoring began.

She was known as only 010 for some six years, before being given her second name, which most people know her by, as part of a competition run by *BBC Wildlife* in the summer of 2011. Other contenders were Sophie and Iona. I think Iona would have been a lovely name and very fitting given that the lovely Isle of Iona is in the Inner Hebrides, but the winning name, as we now know, was Occasus. While not immediately obvious, this name is even more fitting than Iona.

The Latin word *occasus* means 'downfall', 'end' or 'west' and is often used with the word *solis* meaning sun. Therefore, the interpretation of her name is, 'the sun sets in the west'. Clever thinking could have settled for West, short for West Coast Community, but the sun setting is very symbolic, hinting at the final setting, the last glimmer of light before it goes out for this family and we lose them. As we approach the end of this section of the book, I want to take the liberty of pressing home my warning of the dangers our marine life faces.

The West Coast Community have a huge number of pressures and threats, from entanglement and chemical pollution to noise disturbance and physical damage. Each on their own may not necessarily always be lethal, but cumulatively these pressures build, and it is the cumulative stress the orcas are under that is likely the reason behind their demise.

One threat not covered so far is that of climate change, and it's a big one. By now we all know about climate change, or should, and the Hebrides will not avoid its inevitable impacts. As its waters warm, there will be shifts in marine species composition and distribution, as well as the timing of events such as plankton blooms, which in turn impact the rest of the food chain.

There is little evidence, yet, of direct impact on the West Coast Community, but there are indications that the Hebrides are seeing a greater number of common dolphins, with an increasingly northern spread. This is great for whale-watching enthusiasts as common dolphins are gregarious and playful and put on quite a show. However, it isn't great news for species such as the white-beaked dolphin that prefer colder waters, and we are already experiencing reductions in sightings of this species.

A recent study showed that some cetaceans are making pole-ward movements from their home ranges as waters warm, leading to changes in the timing of their migrations. Other species so far seem unaffected, but the reality is that the impacts on cetaceans of rising sea-surface temperature, altered ocean chemistry and reduced sea ice has been scarcely studied. Whether these changes have already impacted the West Coast Community, we don't know, but we can be sure that big changes are underway, both on land and in the sea, and even the top of the food chain will not be left unaffected.

We know that biodiversity is being lost at an alarming rate. The WWF produces a Living Planet Report every two years, which aims to address some of the key issues our wildlife is facing. These reports, bringing together numerous global studies in 2020, showed a fall of over 68 per cent of our wildlife populations in just forty-five years. Where we will be in another forty-five years I dread to think.

We have all seen the photos of skinny polar bears stranded on a bit of broken ice sheet in search of food. Those images are ingrained in most of us, I would imagine, the polar bear being such an iconic creature. So far, it's been all too easy to assume that this plight isn't relevant to the orca as it hasn't been seen as an animal at risk of extinction. With more than 50,000 individuals thought to roam our oceans, it seemed orcas were safe, plentiful and resilient.

The International Union for Conservation of Nature (IUCN) maintains a database of all species and works to identify which are at risk of extinction. Those most in need of help are listed as either Threatened, Endangered or Critically Endangered. However, despite all we know about orcas, they are still listed as Data Deficient and therefore not recognised as a species at risk of extinction. Gradually, we are learning that nothing could be further from the truth.

The West Coast Community is on borrowed time and will soon enough be gone. Some will say that there are only ten of them. Surely nothing to get too worked up about. However, our small population of orcas is only one among many who are living on borrowed time, whose ability to create well-defined ecological niches, often based on the food they eat, is the very thing that is leading to their demise.

On top of this, they are swimming in a cesspool of our harmful chemical legacies and our failure to care that they need a quiet ocean. These things are driving populations across the globe to collapse, even though some individual countries have assigned conservation status to their local population.

In Prince William Sound, Alaska, there are populations of resident and Bigg's (transient) orca. The residents are much like those off Telegraph Cove and the San Juan Islands. They return to Alaska and Prince William Sound year after year in their search for Chinook salmon. The Biggs roam the coastline in search of seals and other marine mammals, frequently come south and have been seen many times off Tofino on the western side of Vancouver Island.

There is another group of transients called the ATIs, who are very special and were once one of the most frequently sighted groups of orcas in Alaska. However, they now face the same fate as our West Coast Community. Back in 1988, a young graduate biologist, Eva Saulitis, started her research in Prince William Sound. While she was tasked with photo-ID work of all humpback whales and orcas, it was the ATIs that she was particularly captivated by. In 1988, there were twenty-two individuals in the population.

The following year, on 24 March 1989, the large oil container *Exxon Valdez* ran aground on Bligh reef in Prince William Sound and started spewing many gallons of crude, black oil into this pristine environment. The recovery attempt was marred by countless errors, including starting too late, use of damaging chemical dispersants and ineffective attempts to contain and clean up the mess.

Over 1,600 squares miles of Prince William Sound was affected. Some 500,000 seabirds and three thousand sea otters died. It was estimated that over half of all wildlife perished in that first year. The oil covered their bodies, leaving them unable to fly or swim properly. They ingested food that was covered in the toxic sludge. Their lungs were filled with burning vapours from the oil.

The day after the spill, the ATIs were seen swimming right by the *Exxon Valdez*, breathing in that toxic air, damaging their lungs.

Exxon Shipping Company spent $2 billion on the clean-up and, a decade after the tragedy, claimed there had been no long-term environmental damage. Conservationists and locals do not agree with them. Two decades later the sea otters and salmon appeared to have recovered, and after nearly thirty years the marine-mammal population as a whole seemed to be following. The same cannot be said for the ATIs.

The impacts of the oil spill were devastating and the loss of even a few individuals from an already small population size means the impacts are more severe. A year after the spill a female, known as AT19 or, more affectionately, 'Berg', was found dead. She was twenty-one years old and in the prime of life, but she had died. She was spotted from a plane, floating upside down in an area that was largely covered in ice with a few channels in it. By the time the researchers arrived to do a basic necropsy, her body had beached and she was already badly decomposed. While a full necropsy wasn't carried out, it was evident from her empty stomach that she had not eaten in days. The cause of death was marked as 'unknown' but with no obvious reasons such as entanglement or gunshot wounds, many believed the oil spill and its toxins were to blame for her untimely death.

The loss of Berg is far greater than one individual. Her death represents all the offspring she will not bear, and all the grandchildren she will not have. The loss of a female means the loss of a matriline and a weakening of the population's ability to recover and grow. Just as the death of Lulu was such a blow to the West Coast Community, so was Berg's to the AT1s.

Within the first year of the spill, nine individuals perished and today only three males and four females survive: Marie (AT2) and her son Ewan (AT3), Chenega (AT9) and her son Mike (AT10), females Paddy (AT4) and Iktua (AT18), who have not had offspring, and the last is the male Egagutak (AT6). With all the females being older than forty years and therefore likely to be post-menopausal, the chances of any new offspring are slim to none. In fact, there has not been a single documented birth in this population since before the *Exxon Valdez* oil spill. Ewan is the youngest, born in 1984 and therefore nearing the end of the average male lifespan. Given their lack of breeding with other orcas, it is just a matter of time until we lose this population forever.

Reading Eva's book *Into Great Silence*, you cannot help but be mesmerised by her obvious deep connection, not just to the AT1s but also to Prince William Sound. Sadly, in January 2016, she lost her battle with cancer. She is remembered by many across the world for her passion for orcas and for nature, and for her eloquent and poetic writing. Only a few in this world are gifted with such a talent.

The AT1s are not the only population to be affected by an oil spill. In 2010, the Deepwater Horizon oil spill had a hugely damaging impact on the Gulf of Mexico population. Birds, turtles, fish and marine mammals were all affected. The orca population here was already very small and on the decline. In 1994 there were 277 individuals, and by 2003 just forty-nine. By 2009, the year before the oil spill, only twenty-eight were left. So, even before the oil spill their numbers had dramatically reduced, possibly due to PCB contamination, but their decline is not fully understood. While no orcas were reported to have died as a result of the oil spill, we know from the AT1s that the impact on reproduction and the long-term outlook for an already small population is not good. Sadly, this population is also set for collapse.

The Strait of Gibraltar is a nine-mile-wide stretch of water that separates two countries and continents: Spain in Europe and Morocco on the African continent. It also connects the Atlantic Ocean to the Mediterranean Sea and it is thought that without the strait, the Mediterranean would evaporate within a thousand years. A huge number of cargo ships and vessels travel through it every single day, as it is the main transport route into and out of the Mediterranean. It is also the summer home of a group of orcas.

The orcas in the Strait of Gibraltar, who are also seen in the Gulf of Cadiz, are seasonal visitors and are the only orcas you will see in the Mediterranean. Thirty-nine individuals have been observed year on year since they were first catalogued in 1999, and can be grouped into five sociable pods that tend to stick together. These groups appear to be consistent from year to year.

Like the West Coast Community, they are genetically and culturally isolated. The Strait of Gibraltar is close to the Canary Islands, where orcas are also seen, and so, for a long time, it was assumed that the two populations were the same. However, it appears that there is absolutely no mixing of the two. The Strait of Gibraltar population does not appear even to visit the Canaries, although we are not yet sure where they spend autumn and winter.

This population of orca returns year after year in the summer months to nurse their young and feed on Atlantic bluefin tuna (*Thunnus thynnus*). As we have seen with other orcas that feed on specialist prey, they have specialised hunting methods that are likely passed from one generation to the next as a learnt behaviour, as part of their culture.

They have two main hunting techniques. One is to simply chase the tuna until exhaustion, which is not an easy task as tuna are basically torpedoes and built for speed. The other method, not popular with fishermen, is to take them from the fishing lines. Whale-watching companies know to follow fishing boats for the best chance of encountering orcas. However, this second strategy isn't used by all five social pods so it isn't clear whether this is an adaptation by just a few groups.

This specialisation is another example of how orca in different parts of the world have developed their own niche prey species. However, it is this group's love of tuna that is causing their numbers to dwindle so alarmingly. Bluefin tuna is a highly prized and desired fish, but has been so heavily exploited in the Mediterranean that it is now listed as an Endangered Species. The world's insatiable desire for sushi, particularly in the Asian market, has led to overfishing, with illegal fishing an additional problem.

Fishermen in the Mediterranean are also moving away from dropline fishing, so the learnt behaviour of taking fish from these lines is less possible. Therefore, there are not only fewer tuna in the Mediterranean, but also fewer opportunities for the orca. Since the fishermen have stopped using droplines, there has been a complete cessation in breeding, which suggests that orcas are no longer in optimum condition for reproduction.

PCBs have also been named as a major risk to the health and viability of this population. The bluefin tuna is a predator fish, consuming mackerel, sardines and herring. They can live for forty years and weigh up to 1,500 pounds, which is mighty big for a fish. This longevity means that there is a long time for PCBs and other pollutants to build up in their flesh, all of which are passed onto the orca.

As we read in Lulu's chapter, these toxins can cause a whole host of problems, including a compromised immune system, cancer and an inability to reproduce. Between this heavy toxic load, a depleted food source and an increasingly noisy habitat the outlook cannot be described as hopeful.

Due to their small population size, their genetic and social isolation and their limited ability to reproduce, the Strait of Gibraltar orcas have been listed as Vulnerable in the Spanish National Catalogue of Endangered Species (Royal Decree 139/2011), but recent studies suggest this categorisation should be heightened to Endangered. Because of this, a marine-mammal protected area has been proposed for them.

The IUCN has a special task force called the Marine Mammal Protected Task Force Area which has identified the Strait of Gibraltar and the neighbouring Sea of Cadiz as an Important Marine Mammal Area where conservation efforts should be focused to help protect this population which is headed for collapse. Hopefully, the special task force will be able to set up a conservation plan, likely concentrated on limiting seasonal underwater noise and improving the tuna stocks, which should help bring the orca population back from the brink.

The Southern Resident orca, or the SRKW, or Orca of the Salish Sea, spends most of its time in the waters between British Columbia in Canada and Washington State in America. Of all the orca groups, it has received the most attention by the media and its struggles are known the world over. From spring to autumn, these orcas spend their time hunting salmon in the Haro Strait and the San Juan Islands.

Split into three pods, called J, K and L, they are all related and part of the J clan. They are often seen hanging out together, either in their own pod, or with members of other pods, but they never associate with orcas not part of their clan. The SRKW do not mix with the resident orcas found further north in British Columbia, such as those up in Telegraph Cove I had twice travelled to see.

Along with the Northern Residents, the Southern is the most intensively studied orca on the planet, and research into this group has paved the way for our understanding of other populations across the globe, which is why they feature so heavily in this book. Sadly, as was mentioned in Floppy Fin's chapter, captures for the entertainment industry started with them.

At one point it was thought that there were more than two hundred individuals. However, in September 2022 the number stood at 73, the lowest the population had been in thirty years. The numbers were low in the 1960s and 1970s due to the live captures of nearly fifty, but after live captures were banned the numbers increased.

Sadly, in recent years, the population has been reducing again. As older members die, the numbers are usually balanced by new members being born. However, the SRKW have not had much luck and in the four years to 2020, only a single calf survived. On top of that, otherwise relatively young adults are dying. It is thought that there are four main reasons for their recent decline:

These orcas feed almost exclusively on Chinook salmon, whose numbers have declined drastically thanks to overfishing.

The use of dams in the Snake River system, which runs from Wyoming into the Columbia River and out in the Pacific Ocean in Washington state, has altered the river system that thousands of Chinook have historically travelled up to spawn. Without healthy rivers, the salmon do not breed and the next generation is never created.

As with the West Coast Community, legacy chemical pollutants such as PCBs are found in high concentrations in these waters. Waterway traffic adds to this pollution.

Finally, noise from cargo ships, cruise ships, fishing boats and, ironically, whale-watching boats have made the Salish Sea a rather noisy place, limiting the the ability of orcas to chat away to each other, to socialise, to hunt and to maintain their close connections.

Ken Balcomb studied these whales for over forty years and knew all too well the struggles they face. Sadly, Ken passed away in December 2022. His Centre for Whale Research based on San Juan Island, with its hundreds of staff and volunteers over the years, continues to work with the community to raise awareness, monitor, report and dive into action to help their plight.

It is people like Ken who, having dedicated his life to these animals, give me hope. There is now a task force taking action to help this population recover. It has been a painful process, as it usually is in politics when no one wants to foot the bill or make unpopular changes. Money has been promised on both sides of the Canada/USA border and controls have been put in place, such as extending whale-watching distance limits and the banning of fishing in key areas. However, it is doubtful that these measures will be sufficient.

The SRKW are now seen less and less in the Salish Sea as they expand their home range in search of prey. When I volunteered with the Strawberry Isle Marine Research Society on the west coast of Vancouver Island, I examined their 25-year data set. Historically, they are only seen a handful of times each year, but recently had been seen in the summer months more frequently. As their dependable food supply in the Salish Sea disappears, they are venturing further away.

Fortunately, the First Nation people of Tofino and their neighbours are keen to support any visiting orca and look after the natural world. River restoration days for salmon runs are a frequent event in Tofino. During these events, the community rallies together to plant native vegetation along the river in key areas, not only helping to keep the rivers cool and improve biodiversity, but also helping prevent sediment and pollutants from building up. This in turn helps restore the river to a condition suitable for the salmon to spawn in.

To lose the SRKW would be a tragedy and a travesty. They are the most loved, most viewed and most studied orcas on the planet. We know why they are struggling and failing to thrive, but action is coming too late and too little and, sadly, this is because of mankind's greed and need to dominate and exploit the planet's natural resources. We need to learn to share better and understand that the oceans can be plentiful only if we look after them.

Other populations that have been highlighted at risk of collapse are those found in Japan and in Brazil. Again, it is those high concentrations of PCBs and chemical pollutants that present the greatest risks. This will be true of any coast that is highly populated and industrialised.

A paper published in the prestigious *Science* journal in late 2018 highlights how real and devastating the outlook is for the orca. So much so that it has been dubbed an 'orca apocalypse', with 50 per cent of the global orca population threatened in the next hundred years, and this only takes into account the effects of PCBs. The collapse of fisheries, climate change, the captive trade, entanglement in fishing gear and noisy waterways all add to this risk.

Aside from the big events such as oil spills, which are responsible for the decline of the ATls and those in the Gulf of Mexico, the single biggest threat appears to be the legacy of PCBs in our environment. As we saw with Lulu, they pose a real threat to the West Coast Community. She had levels higher than had been previously recorded, showing that orcas don't necessarily have to live in heavily industrialised areas to suffer. Feeding at high trophic levels is enough to create a deadly contamination level.

The incredibly scary thing is that, while PCBs have been banned since the 1970s, only about 20 per cent of land-based PCB-containing products has been disposed of. Eighty per cent is still to be dealt with. Leaving it in the ground is not an option as the chemicals will leach into the soil and find their way to the sea. So how do we go about proper disposal? By either treating PCBs with another chemical or incinerating on land, neither of which are ideal solutions.

The United Kingdom accounted for only about 5 per cent of global PCB production, whereas the US accounted for about 50 per cent. Russia, Germany and France were the next biggest producers.

Although the manufacturing of PCBs has ceased, some

products are still in circulation and disposal remains an issue. Oceans do not keep to country boundaries. What happens in one country has the potential to affect marine life on the other side of the planet. This is a global problem, and we must unite to address it.

This has been a rather bleak chapter, but it is time to face the harsh reality of what we are doing to our planet and the devastating impacts it has on our marine life. Most people will be unaware that the iconic and robust orca is facing such a dismal future, or that it is a direct result of man's actions and our lack of understanding or consideration of nature. The science has been available for a long time, but we just haven't wanted to hear it. It is not too late to change things. We can and must do better, and hopefully we can reverse this decline and restore our oceans to their former glory.

MONEYPENNY

Protecting 007

MONEYPENNY (007) WAS FIRST identified in 2004, quite some years after the first group was seen in the 1990s. She was not the last to be added to the West Coast Community catalogue, but seventh, as you might have noticed. It appears that there are two groups of orcas within the West Coast Community: the original group catalogued in the early 1990s (John Coe, Floppy Fin, Nicola, Comet, Moon and Lulu) and a second, which was added to the catalogue almost ten years later (Moneypenny and Aquarius in 2004, Puffin in 2000 and Occasus in 2005). All except Aquarius are female and it is possible that they are younger than the originals.

When I shared this thought with the real expert on this group, Dr Andy Foote, he was not so sure. He thought that it is rather more likely that they simply weren't seen together at the same time. As we have learnt in this book, it is hard to track individuals and they are rarely seen as a whole group. It is possible that Moneypenny is around twenty-seven years old with, potentially, another fifteen years to reproduce. In that case, it would be likely that she has been sexually mature since 2004 and we would have hoped to have seen her with a calf. She has been spotted several times since, but never with a calf, and it is assumed that she has not been successful in breeding.

Orca mothers and calves are extremely close. The calves will rarely stray, often swimming in synchrony or close physical contact. Scientists at the Marine Mammal Research Centre at Vancouver

Aquarium have used drones to take aerial photos of orcas in the wild, and I spent several months looking through thousands of these images cataloguing individuals. Photos were taken in rapid succession through a flight time of about ten minutes. Batteries were then changed, and another flight made. Clicking through the photos rapidly is like watching a film, but you can stop on a particular image and zoom in. I often did.

It is a thing of beauty to watch mother and calf together. The youngster will spend much of the time swimming directly under-neath between its mothers' belly and tail, which is known as eche-lon swimming. Riding in her slipstream, the calf rarely has to make much effort to keep up, or to suckle at her mammalian slits.

It is thought that up to 63 per cent of pregnancies are unsuc-cessful with the calf lost late in pregnancy or shortly after. Of the calves who are actually born, many have a tough time in their first year. In fact, up to 50 per cent of neonates do not survive to their first birthday. Given the huge amount of time and energy invested into their nineteen-month pregnancy, losing half of all babies is tough for the mothers.

It is not really known why so many die. Females in resident pods appear to be excellent mothers, nudging their babies to the surface to breathe and often letting them ride on their back so they can breathe easily. Aunties and grannies stand by to help, and even uncles have been seen to help out. It appears that the calves have a lot of support, but still suffer very high death rates. With several years between each sighting of Moneypenny, it is possible that she has had a baby who died in the interim. Maternal behav-iour is thought to be at least partially learnt so, with no other family members having babies, it could be that Moneypenny has not learnt the skills she needs.

More likely though, and if Lulu is anything to go by, we can assume that incredibly high levels of PCBs in Moneypenny's body, or simply being so inbred, have prevented her from conceiving. She hasn't been seen in many years though and is now thought to have passed on. Another female lost, and with her, the chance of more offspring.

Moneypenny was the seventh orca added to this group so her catalogued number is 007. If she had been a male, I am pretty certain she would have been called James Bond but, as a female, the next best name was Moneypenny. While I have to admit to not really being a fan of the James Bond films, I have seen enough to know that his job is to do all he can to protect those he is assigned to. But now the tide has turned, and we must protect 007. What better way to end this section of the story of the West Coast Community than to look at ways we can help them and other orcas across the globe.

I started this book by talking about my first visit to Telegraph Cove on Vancouver Island when, in 2012, I kayaked with Adam in the hope of seeing the Northern Resident orca that Erich Hoyt had described in his book *Orca: The Whale Called Killer*. Sadly, on that visit, it wasn't meant to be and I left without seeing them.

In 2016, after I had finished my visiting scientist post at the Marine Mammal Research Centre in Vancouver, we travelled once again to Telegraph Cove, where we hired an old camper van and took a few days to travel up to the northern end of Vancouver Island, camping here and there by the beach for the night. It was my version of perfection; a simple dinner cooked on a tiny stove, enjoying a cool beer as we watched the sunset, wrapping ourselves in blankets as day gave way to night.

Telegraph Cove hadn't changed a bit, which was comforting. I was booked onto a whale-watching boat trip in a few days' time, so we decided that we would try our luck at spotting them from kayaks and land again and joined an early-morning kayaking tour. The mist from the forest and ocean hovered over the water and clung to the trees that grow all the way up to the edge of the rocky outcrops, marking the end of the land world and start of the ocean.

We could hear lots of puffs from small marine mammals and, sure enough, as they emerged from the mist, we were joined by a dozen or so Dall's porpoises. They are more gregarious than the harbour porpoises seen from the *Silurian,* but there was still no breaching behaviour. They just milled around us curiously, enjoying their start to the day as much as we did. Incredibly beautiful, they are almost like mini orcas with their black bodies and the flashes of white from their belly and dorsal fin, but they were the closest I came to seeing an orca that morning. Still, it was an exquisite way to start our day.

In the afternoon we took a short hike to a viewpoint over Johnstone Strait on the first section of the Blinkhorn trail. We would need a day for the full trail, but this took us only about an hour and the views were spectacular. Vistas out over the islands that make up Broughton archipelago with their ancient rainforests surrounded by the deep blue seas and kelp forests. We had the viewpoint to ourselves for the first hour and both stood to attention, keeping a lookout for that telltale blow or dorsal fin, and this time we were in luck. A pod of orcas was moving slowly and sociably south down Johnstone Strait. They were a bit far away to get any good photos or to identify them, so we just enjoyed watching them as they went about their day.

Soon enough we were joined by a couple of hikers and it was a proud moment as I watched my husband pointing the orcas out to our new friends and telling them facts and stories about the Northern Resident population. I sometimes wonder whether I bore him senseless when I talk to him about these things, so it was special to see that he had been taking it all in and seemed as excited as I was. Some of the facts were slightly wrong (an orca cannot swim at 100kph), but he was so enthusiastic I didn't have the heart to correct him.

The next day we set off early and hiked to the Blinkhorn Peninsula, taking a good few hours to complete the walk. We were ready to spend the entire day there so, along with our packed lunches, plenty of water, sunscreen and binoculars, Adam insisted we also take one of our fold-up chairs. It didn't seem like the smartest idea while we were hiking in the heat, climbing up rope ladders and scrambling over boulders and fallen trees, but it certainly seemed like a good idea once we were there. Our hike was worth every drop of sweat.

We didn't have to wait long until we saw the first pod of orcas. Like the day before, they were heading south to the Robson Bight Ecological Reserve, where Michael Bigg, Erich Hoyt and Graeme Ellis made the discovery of the rubbing beach. This area is of huge cultural importance to the orcas here, who have for generations gone there to socialise, play and mate.

When Bigg and Co made this discovery, they worked hard to ensure the area achieved protected status. Logging remains, sadly, big business on Vancouver Island given how heavily forested it is with large mature trees. Removing these trees is hugely damaging. The treeless ground becomes unstable, and loose mud and silt get washed into the rivers, clogging the pristine environment, making it unsuitable for the spawning salmon which are key to the ecosystem both on land and in the water. This extra silt reduces both water quality and the amount of food available for marine life, eventually reaching through the whole food chain.

Jim Borrowman, a local marine-mammal expert and whale-watching captain, along with Erich and several others, lobbied to have this rubbing beach and the area surrounding it, both land and sea, protected to ensure the Northern Resident orca would always have a safe place to return to. They understood how important it was to their culture and way of life, and that it was imperative that it was saved. Their efforts paid off and the Robson Bight (Michael Bigg) Ecological Reserve and the surrounding forest and rivers are now a legally designated sanctuary for orcas under British Columbia provincial legislation and under the Species at Risk Act (SARA), a Canadian federal legislation.

While commercial fisheries are allowed in the sanctuary during the salmon season from July to November, no public boaters, motorised or not, are allowed to enter the marine part and no trees are to be removed from the forest. Permits are required for anyone wanting to carry out research or education. The inclusion of the forest and rivers is an important step in the acceptance that the various components of the ecosystem here are tightly connected, and to protect one part, you must protect the whole ecosystem.

This strict legislation means, on paper at least, that there is a refuge for the Northern Residents where they are free to live their lives as they have done since before man started to interfere. On the whole this is a triumph. There are designated wardens, such as Jim Borrowman, who monitor the reserve to ensure the rules are respected. Enforcement and co-operation are imperative to the success of a marine protected area. Without them, it can be nothing but a paper park. But getting sites designated is a huge step in the right direction to affording protection to these animals.

Marine protected areas (MPA) like the one at the Robson Bight Ecological Reserve are a powerful conservation strategy that can help us safeguard the future of a particular community or species or even whole ecosystems. Most countries have signed the United Nations Convention on Biological Diversity, which aims to protect at least 10 per cent of the ocean's surface by 2020. This target was missed, with only 7.5 per cent declared as MPAs by that year.

However, since whales will rarely stay within such confines, we would actually need to protect about 30 per cent of the ocean's surface to properly accommodate them, as well as continuing with whaling bans. The UK has committed to protecting 30 per cent of our oceans by 2030. This ambitious target requires big action.

In 2019, the Scottish Government announced the establishment of the Inner Hebrides and Minches Special Area of Conservation (SAC) for the harbour porpoise, which covers much of the west coast of Scotland and is one of the largest protected areas for the harbour porpoise in Europe. Like all other cetaceans, they are already listed as a European Protected Species, which means it is illegal to intentionally harm or kill one. The ultimate aim of the SAC is to keep the harbour porpoise population here at the same favourable level it is currently. This region has a higher-than-average abundance of the species and is known to be an important area for calving, especially in the summer months.

This SAC plans to offer extra protection through research, monitoring potential threats such as underwater noise and impacts of climate change and also management of the fisheries in favour of prey species. Things that are to be addressed include the use of ADDs in aquaculture which, as we have seen in Puffin's chapter, can generate a lot of underwater noise, pollution that is harmful to marine mammals such as harbour porpoises, and many other species. The plan also recommends the banning of bottom trawling and drift nets, which damage the seabed and therefore impact on their preferred prey species, the sand eel. Additionally, vessel routes and speed restrictions might be considered to help minimise collisions. As we saw in Moon's chapter, in 2015 a male orca, possibly Moon, was found dead on a beach on South Uist. He had several broken ribs and a broken jaw, and it is thought that he died as a result of a collision with a boat. By slowing boats down, the chances of collisions will be reduced as will the risk of death if a collision occurs.

This SAC covers a huge area, some 13,800 hectares, but despite its size, more protected areas were needed to reach the 30 per cent target and weighing up where to protect and for what species, taking into consideration industry needs and politics, it is not a simple decision.

The data collected by volunteers on the *Silurian* and the community sightings reported to the HWDT feed into the body of knowledge about where marine protected areas are most needed and where they would have the most impact. Together with the Whale and Dolphin Conservation (WDC), the HWDT has worked and continues to work tirelessly to lobby for increased protection for the marine mammals in our waters.

The rewards of this work were seen when the Scottish Government announced four new MPAs in Scottish waters in December 2020. These were first proposed back in 2011 by WDC, but continued pressure and a building scientific body of evidence for their value, along with the support of 36,000 signatures, led to these four new designated areas.

Three of these new protected areas are on the west coast of Scotland, and of these three, two are designated for marine mammals: the Northwest Lewis MPA and the Sea of Hebrides MPA.

The Northwest Lewis MPA is in the Outer Hebrides and designated to protect both Risso's dolphins and sand eel communities on the seabed. Sarah Dolman, Mike Tetley and Nicola Hodgins from the Whale and Dolphin Conservation have been key to the research to get this designation. For over a decade, each summer, they have been working to photo-ID and study the Risso's dolphins off Lewis and gain a deeper understanding of the importance of this region to them. We didn't see any Risso's dolphins on our trip – as they are so distinctive, we are confident that we didn't just mistake them for another species. They have a big blunt head with no rostrum, or beaklike projection, at the front of their head. Their most distinctive feature though is their skin, grey in colour but covered in white scars that they accumulate over life. These scars predominately come from the food they eat: squid!

Risso's dolphins love squid and spend a lot of their time diving into deep water to find them. This is one of the reasons that they aren't seen very much. They are usually deep underwater and, even if they are near the surface, they are unlikely to approach a boat as they are rather shy. But if you are super lucky, you might get to see them in the huge pods, of up to two hundred individuals, that they sometimes gather in. What a sight that must be.

While it wasn't meant to be for us, there have been enough sightings over the years to know that the Hebrides are an important area for them, particularly the Outer Hebrides. By the continental shelf, the water gets deep very quickly, and they can get to the squid which live at depth. However, they are often seen near the Isle of Lewis in much shallower water. It is here that they come to breed, give birth and nurse their young. It is definitely an area worth protecting.

A real concern for Risso's dolphins is the use of sonar, which gets tested during the biannual Joint Warrior Exercises in the Hebrides. Deepwater species like the Risso's are especially susceptible to being harmed by underwater sonar, as they may ascend too rapidly and get the bends. Mass stranding of marine mammals that feed at depth are well documented in response to sonar testing. Hopefully, this new MPA will offer them greater protection from this acoustic threat.

Provided we can create a safe environment for Risso's dolphins, we might well see an increase in their numbers in the coming years. Typically, they prefer warm water, so with climate change and a warming ocean, perhaps more will venture further north into the Hebrides to raise their young.

The Sea of the Hebrides MPA includes the Small Isles, or as we knew them on our trip, the Minke Triangle. It is an area that has long had the reputation as a hot spot for minke whales and so as we approached the Isle of Rum on our first day at sea and then again on our second to last day, we were all hopeful. We didn't get to see any minke whales around Rum, but we did see a total of seven minke whales in our trip. That isn't bad going. Plus, as sightings of minke whales go, we had a pretty good one with our playful mate, who stayed with us, displaying his big white belly for half an hour. Knowing that this area is now protected to help ensure the safety of this species is fantastic.

This area is protected for another of our ocean's gentle giants, the basking shark. As the world's second largest fish, the basking shark can reach ten metres in length and comes to the Sea of the Hebrides in great numbers in the summer months to feed on the rich plankton blooms that occur in this highly productive region. It is also postulated that the Sea of the Hebrides is an important mating and courtship area for the basking shark. It is the only MPA in the world dedicated to protecting this species.

Sadly, on our trip around the Hebrides we didn't see a single basking shark, which was a little disappointing. No doubt they were around as they are usually present between June and October. Perhaps they were just filter-feeding under the water and so not visible at the surface when we were near. I mustn't grumble though, we were more than spoilt with the incredible sightings we did have. I will just have to venture back another time for an encounter with basking sharks and Risso's dolphins.

These new MPAs are a huge step forward in safeguarding the marine life around our coast but it has taken a colossal amount of work from countless dedicated people to get here. With these new MPAs, the UK has not only reached but surpassed its target, with 38 per cent of UK waters now having some level of protection.

While it is unquestionably a major celebration to have these new sites to protect the minke whale, Risso's dolphin and the basking shark, it does beg the question as to why no MPA has been proposed to help protect the West Coast Community. This despite their importance, not just as a keystone predator and possibly the last remaining individuals of an entire ecotype, but also as one of the most iconic and beloved marine mammals on the planet.

In 2013, a study was published that demonstrated that the West Coast Community should be considered as a separate Population Unit, also recording that there are, or were, in fact only ten individuals remaining and that they require separate conservation management.

Eleven years on, nothing further has happened and, with just two elderly bulls thought to remain, it could be too late. I imagine the reason is not just because we know very little about them, but also because very few people know about them at all. If more people were aware of this population and their importance, their significance and their uniqueness, there would be public outcry that more was not being done to protect them.

While the West Coast Community is dwindling, the orcas of Shetland and Orkney appear to be thriving. With social media, more and more people are becoming aware of them and heading to the Northern Isles in ever greater numbers to see them. There are many individuals that are known fondly by locals, some that appear to be resident of the Northern Isles, such as Razor and Busta, and others that are summer Viking visitors from Iceland, such as Gunnar, Brèagha, Summer, and Mousa and her offspring Tide. Land-based watching is incredibly popular here, with the orcas coming close to shore to hunt seals. This allows great opportunities to view them with no impact at all. Long may this continue.

We must ensure that we protect these orcas and I believe that an MPA around Shetland would be beneficial. However, for our summer visitors, their time around Shetland is only half the story. Many spend their winter in Iceland, and protecting only half of their territorial range may not be sufficient. While there is a lot of collaboration between scientists in Iceland and Scotland who research this population, could we achieve more through politics and legislation?

For this population to continue to thrive there are multiple factors to consider. We need to ensure that the herring stocks stay healthy in Iceland, which means monitoring the impacts of climate change on this species and managing the fisheries to keep a sustainable catch allocation. The seal population around Shetland must remain large enough in the summer months to sustain a growing orca community. There needs to be safe passage between Iceland and the Northern Isles to prevent collisions between vessels and orcas, and also safe boating practices around Iceland, Shetland and Orkney to avoid collisions and disturbance.

There is a respectful and growing whale-watching community in Shetland. Each year, more tourists are drawn there in the summer to try their luck at seeing the orca. It is wonderful to see more people be interested in learning more and seeing them in the wild. We must ensure best practices of wildlife watching continue in a growing market. For the West Coast Community the outlook is bleak, but let us learn from them and do better to protect others. We are incredibly blessed with abundant marine life around the British Isles, including plenty of the ocean's most iconic and impressive species. The mighty orcas that we see in our waters, whether they be John Coe, Aquarius, Razor, Mousa or any of the others that visit our waters each summer, all deserve our protection. Knowledge is power and we must harness this to protect the remaining individuals that need our help.

PART TWO

The Voyage of the *Silurian*

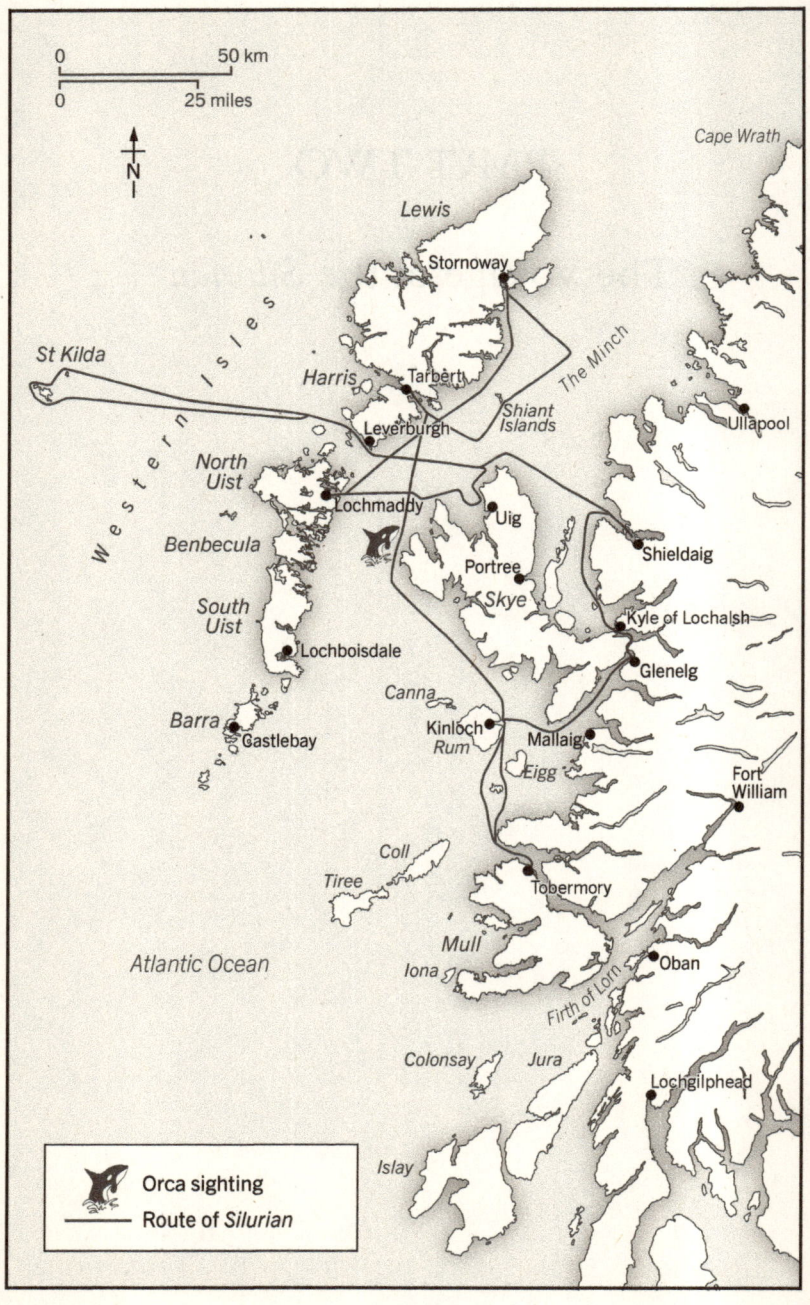

0
50 km

0
25 miles

N

Cape Wrath

Lewis

Stornoway

St Kilda

W e s t e r n   I s l e s

Harris
Tarbert

The Minch

Leverburgh

Shiant Islands

Ullapool

North Uist

Lochmaddy

Uig

Shieldaig

Benbecula

Portree

Skye

Kyle of Lochalsh

South Uist

Lochboisdale

Glenelg

Canna

Barra
Castlebay

Kinloch
Rum

Mallaig

Eigg

Fort William

Coll

Tiree

Tobermory

Atlantic Ocean

Mull

Iona

Oban

Firth of Lorn

Colonsay

Jura

Lochgilphead

Islay

Orca sighting

Route of *Silurian*

TRAVELLING TO TOBERMORY

Monday 1 September 2014

Journey: Heathrow to Tobermory, Isle of Mull
(Distance: approximately 520 miles)

IT WAS THE BOOK *Orca: The Whale Called Killer* by Erich Hoyt that ignited my desire to research the orca but my passion for the marine world has always been there. My obsession with marine mammals started at a very young age. While my peers hung posters of boy bands from their bedroom walls, my walls were covered in pictures of leaping dolphins. My room was cluttered with trinkets and figurines collected over years as gifts from family who knew my interests all too well.

However, my passion started even earlier than that. Possibly before I could even walk or talk, or at least this is what my mother believes.

I was born with congenital heart disease. The wall that separates the right from left, the oxygenated blood from the deoxygenated, simply didn't form when I was in the womb. So I spent a great deal of time at Hammersmith Hospital in London, fighting off infections and colds, my poor parents having to wait until I was considered big and strong enough to withstand the surgery, then still novel.

At Hammersmith Hospital, there is a tunnel that connects the old imposing Victorian building from the modern, but slightly soulless, new building. The tunnel starts at ground level, and painted onto the walls on both sides was the shoreline lapping at your ankles and sandy beach all around you. As you descend the tunnel, you become immersed in the water, first up to your knees, waist, then shoulders until you are fully submerged. Dolphins play all around you, swimming in between boulders and rocks, their happy faces with their perpetual smile looking directly at you.

To an adult maybe this wouldn't have been that impressive, but as a little girl I used to feel completely at one down there. It was my happy place in what was an otherwise difficult time. I used to look forward to my hospital visits as I grew up so that I could walk through the tunnel and be with 'my' dolphins. I have been walking that tunnel at least twice a year all my life, until I moved out to the West Country when I had my own kids. For over thirty years, that tunnel has comforted me in times of distress, made me smile and given me a sense of stillness and perspective.

As the years passed, the tunnel and its paint showed the signs of many years of wear and bed trolleys being bashed into it. It has now been painted over in beige. I guess it is easier and cheaper to maintain. Still, on my last visit one small section remained of my dolphins. Enough to put a smile on my face but not enough to feel fully immersed in their world. Of course, these dolphins are not *my* dolphins, and the decor decisions taken by the NHS do not revolve around me and my needs, but it's sometimes nice to let your imagination run wild like it does so much as a child.

So yes, my mother might be right. The images and experiences we are exposed to at an early life stage can indeed have long-lasting impacts.

It was only logical and expected then that I decided to study marine biology at university, and I have since been lucky enough to travel to many exotic places with my studies and research, from Papua New Guinea to the Great Barrier Reef, Indonesia, the Black

Sea and of course, British Columbia in Canada, where I spent several months learning all I could about the Resident and Bigg's orcas.

But without a doubt, one of my most memorable times at sea was in the Hebrides when I boarded the *Silurian* in search of the West Coast Community orcas.

My journey to the Hebrides only really began after a short flight from London Heathrow to Glasgow. After a bus ride to Glasgow Central Station, I boarded the train to Oban in the late afternoon for what proved to be one of the prettiest train rides I have ever enjoyed, travelling through the famous Trossachs National Park, passing mountains with evocative Gaelic names such as Beinn Udlaidh.

The journey took us through dense spruce forests, past mountains shrouded in mist, and there was something so calming and peaceful about it, a million miles from the chaos of Heathrow just a few hours before. As I approached the western shores of Scotland, the sun was setting gloriously.

Between the forest trees, you get glimpses of the many, vast lochs that are scattered throughout Scotland: the River Clyde out of Glasgow leading to the Gare Loch, Loch Long and the most famous and largest, Loch Lomond, then Loch Awe before the final stretch of track along Loch Etive. Some are lakes, cut off from the sea in the last ice age, others are sea inlets. As the mountains gave way and the land levelled out, vast swathes of purple heather carpeted the ground.

After a long day of travelling, I was ready for bed but, as I checked into a simple but perfectly adequate B&B to catch my last night of sleep on dry land for a few weeks, my head was heavy with a developing cold.

Early the next morning, I woke up full of it, which is not what you want when you are about to embark upon a ten-day trip in close quarters with eight other people. On my way to the ferry terminal, I loaded up on cold medicine and a big bag of oranges, which I ate rapidly in the hope that the vitamin C would halt the germs in their tracks.

My sea adventure began as I boarded the CalMac ferry to sail to Craignure on the Isle of Mull. It was just a short journey of forty-five minutes, but I of course spent the whole time on the top deck scanning the waters for signs of life, practising my skills of observation.

Mull is an important stop for any wildlife enthusiast. The white-tailed eagles that were reintroduced to Scotland can often be seen on the island, along with golden eagles, otters, whales, dolphins and basking sharks. The harbour is full of fishing boats, but wildlife-watching boats are on the rise as more people become aware of the abundant wildlife of the Hebrides.

Tobermory is one of the main gateways to the Inner and Outer Hebrides. Like something out of a picture book, it is, in fact, the inspiration and setting for the children's TV programme *Balamory*. The main street hugs the shore, and the buildings are painted in bright bold colours. It is the quintessential fishing town.

As I wandered up and down, looking into the windows of quaint shops and tea rooms, I felt very much as if time had shifted and was now running on a different scale. I could feel myself breathing more slowly, deeper, and my body relaxing despite my heavy backpack containing everything I would need for the next two weeks.

I spotted two other lost souls with large backpacks along my meanderings and tentatively enquired as to whether they were headed for the *Silurian*. It turned out they were and, after introductions, Sara, Gemma and I decided that, given we had another hour to kill before boarding, the best thing to do would be to get a drink in the local, Macgochans.

As the time approached, the other five volunteer surveyors, Sara and Gemma whom I had already met plus Karen, Kate and

Kirsty, and I waited on the pier to be joined by the team that we would be working with: Kerry, the marine biologist, Tim, the skipper, and Tom, the first mate.

We were a real mix. Sara was Scottish, from Inverness, but lived in Devon with her friend Gemma, who also joined us. They were both Passive Acoustic Monitoring (PAM) operators, who spent their days (or nights if on night shift) listening on hydrophones placed near oil rigs around the world. Their job was to detect marine mammals on the hydrophone from the whistles, clicks and echolocations they emit, and notify that oil rig with a request to stop drilling.

It is illegal to drill into the seafloor when there is a marine mammal nearby as the noise generated can be highly damaging to them, and oil rigs should also have visual spotters on site to confirm any reports. Stopping drilling is a huge cost, so it is important that these people are skilled. They were both excited to have the opportunity to actually see marine mammals and not just hear them.

Sara had dreams of working out at sea but suffered from seasickness and both she and Gemma were popping seasickness tablets before we even got on the boat which was moored up for the night in the quiet harbour. It wasn't the best sign of things to come!

Karen was a Master's student at Edinburgh University studying Animal Biology. She proved to be the joker of the group with her broad Scottish accent and dry sense of humour. She didn't take life too seriously and we could count on her to lighten the mood and crack us all up.

She grew up on a farm and, while she loved marine life, was most at home in wide open fields jumping hay bales, or on the ice rink where she played curling competitively. In my mind she was the epitome of a Scottish lass, with a head of long red hair, a penchant for the outdoors and a wicked sense of humour. I instantly liked Karen.

Kirsty was also doing her Master's degree and I very much enjoyed hearing about her plans for her thesis. She was infectiously passionate and enthusiastic.

Kirsty also really suffered with seasickness. Too ill to work, she would curl up at the stern, often leaning over the side. Thankfully Sara and Gemma had brought a surplus of seasickness tables, which they generously shared with Kirsty. As the days wore on her seasickness began to subside and she managed to pitch in more, which was a welcome relief to the rest of the team who had to step in for her rotation and thus sacrifice their break and a chance of a warm cup of tea.

Kate, a quirky, adventurous retiree, was the oldest and, I suspect, had not been expecting to spend the next two weeks with such a youthful group of people predominantly in their twenties. We shared a room. She wasn't one to talk too much but it was clear she was no stranger to working holidays and getting stuck in for a good cause. She looked very much at home on the boat and at sea, and the two of us, along with Karen, managed to keep the work going while Kirsty, Sara and Gemma found their sea legs.

Kerry was our marine biologist for the trip and had several voyages on the *Silurian* under her belt plus many research trips with other companies across the globe. She was very knowledgeable, down to earth and somehow made the most incredible cakes in the tiny galley.

Tom was the first mate and general go-to for everything on our trip. If the motor broke, Tom fixed it. If we got seasick, Tom knew the best place on the boat to rest. If we needed food, Tom knew how to throw together a hearty and delicious meal. And much more besides. He was born and raised on the Isle of

Mull and looked like he could be the campaign cover for wholesome outdoor living.

Tim was our skipper, a real outdoor adrenaline junkie with a passion for mountain biking, sailing and kayaking. He was a prankster but it was also evident that he valued quiet time and was more than happy to sit on his own or in quiet conversation at the helm. When it came to discussing protocols and safety on the boat, he was incredibly serious and thorough. But at all other times he was relaxed, approachable and funny.

That was us, our team for the next ten days. A group of nine strangers but, living in very close quarters, we weren't strangers for long.

That first night we had dinner and an introductory lecture on our roles, the wildlife and the local area before it was time to get our heads down. Anticipating an early departure the next day, I lay in my bunk bed with my bag at the bottom and my legs curled up, feeling not only incredibly lucky to be in such a beautiful place, but also incredibly lucky that I am only 5ft 2in tall.

THE ISLE OF RUM AND FINDING OUR SEA LEGS

Tuesday 2 September 2014

Journey: Isle of Mull to the Isle of Rum, Inner Hebrides
(Distance: 50 miles)

ON TUESDAY MORNING WE WOKE early to make our final preparations before leaving Tobermory.

Thoroughly briefed on the protocol for marine-mammal sightings, we didn't have to wait long, although the first few sightings were shambolic. In our excitement at seeing our first harbour seal we forgot the details we were supposed to be recording, who was meant to be doing what, and didn't communicate effectively. Needless to say, we felt pretty embarrassed but soon determined to do better at the next sighting whenever it might come. Over the course of the day we motored some fifty nautical miles but saw little else of note.

Most of the team suffered from seasickness and, one by one, donated their lunch to the sea. Ginger biscuits were consumed with as much speed as could be mustered by those suffering and, as the day progressed, the sickness abated for most. Thankfully I wasn't affected, which meant that, whenever I could, I took the opportunity to chat with Tim the skipper and learn all I could from him. Tim lives with his family in the Outer Hebrides and knows these waters like the back of his hand.

As we left the Ardnamurchan Lighthouse, to starboard, he told me that this was a grand place to spot orcas and that a high proportion of HWDT sightings come from around this area. Ardnamurchan Point is the most westerly location on the British mainland and there is a visitor centre at the lighthouse, commanding views over the islands. It attracts many visitors each year between April and October so, with so many eyes on the water at this specific spot, it is no wonder that a lot of sightings are made here. Also the waters around Mull are deep and fast-flowing, which results in a lot of eddies and upwelling, creating a nutrient-rich area which attracts lots of fish and therefore a lot of seals and porpoises. They in turn attract orcas.

As we passed the Isle of Eigg, Tim talked about what locals call 'the Minke Triangle'. The area between the Isles of Eigg, Muck and Rum is a known hotspot for minke whales. Along with the Isle of Canna, these islands make up the Small Isles, but that day was not my day to see a minke whale. We pulled into Loch Scresort on the Isle of Rum to anchor for the night, ending our first day of surveying and along with it the hope of seeing an orca or a minke.

We pulled up alongside the ferry terminal and disembarked, and all of us volunteers took a stroll along the single-track gravel road that runs from the ferry 'terminal' to the 'centre'.

I use these terms loosely as the 'terminal' is just a slipway with a few huts, and the 'centre' is basically just a visitor centre, a lookout, a small shop, post office, campsite and a few houses. While it is the largest of the Small Isles, it is still only about forty square miles in area. Most of the island is heathland or mountains and only about thirty people live there. However, it is one of the longest inhabited parts of Scotland, with evidence of human settlements dating back to the 8th millennium BC.

It was very pleasant, visually at least, with the sea on our right as we walked under dappled sunlight from the trees that lined the road before disappearing at the foot of the mountain. We chatted away happily as we explored, no signs of life to be seen. It was so quiet and peaceful ... until the midges descended. They seem to have a particular taste for some people, and I was on the menu.

Midges are possibly the biggest downside to Scotland, and especially voracious in the Highlands and Islands. I had been warned about how irritating and persistent they are. I hadn't noticed them when we were on the Isle of Mull or when at sea, where the sea breeze keeps them at bay, but here on the Isle of Rum, with so much wet grass and heathland, they were everywhere.

Battling on as best we could, at the end of the road we arrived at Kinloch Castle. It was once owned by an eccentric family, the Bulloughs, who used it as a sort of party-palace full of extravagant and opulent objects. Built of red sandstone imported from Annan on the Scottish mainland, some 250 miles away, rather than the granite that naturally forms the island, it stands out proud and visible. Never designed to blend in with its environment, it is unapologetic in its grandeur. Attendance at one of the Bullough parties was by invitation only, and unwanted visitors arriving by sea were kept at bay by cannon fire.

When the family decided it was time to go, they gifted the castle to what is now known as NatureScot, leaving behind their fine Edwardian furniture, impressive musical instrument collection and a cellar full of wine. It is now a popular tourist attraction and visitors are made welcome. However, the sun was setting when we arrived and it was closed, so we just peered through windows and took in the grounds before making our way back to the boat.

Most of the people who live on Rum are employees and their families, either working at the castle or looking after the large deer population, which is the subject of the longest running research project on large mammals on the planet. There isn't a lot to do on the island but there is a small school for the few children, and a village hall. I couldn't help feeling conflicted for the children. While they have a whole island to explore and will grow up with a deep connection to nature and the environment, I pondered on what they might miss with so few others to play with.

Perhaps the slower pace will help them grow into confident adults who can resist the temptations of modern life and find more fulfilment than the average person. I hope so.

Back home on the *Silurian*, we were grateful to escape the midges. A peaceful evening followed, watching the sun go down and the dark night set in, and it was beautiful to see day give way to a night sky full of stars. With the gentle lapping of the sea against the side of the boat, and its rocking motion, I fell asleep quickly and deeply.

THE LITTLE MINCH AND
MY ORCA ENCOUNTER

Wednesday 3 September 2014

Isle of Rum to Isle of Harris
(Distance: 70 miles)

THE DAY STARTED OUT pretty uninspiring; overcast but thankfully
not raining. The sea and the sky were dull grey, making it hard to
tell where one started and the other finished.

As I got out of my warm sleeping bag, the cold air hit me. I
slipped straight into a long-sleeved, thermal top and trousers for
breakfast in the saloon. The day's work beckoned so, before heading out on deck, I added the all-important weather-resistant outer
layer: thick waterproof trousers with braces and a matching, red
fleece-lined jacket topped by a woolly hat and gloves with grasp-
able fingertips, thick socks, waterproof boots and a life vest in case
I fell overboard. With or without the life vest, falling overboard
would have been very inadvisable; with strong currents and deep,
cold water, you wouldn't last long before hypothermia set in.
Drowning is a real risk. Before putting on all these layers, a trip to
the head (otherwise known as the bathroom) is wise as it takes a
good ten minutes to get it all off again. No point being caught
short.

The ensemble was topped by a pair of polarised and nonglare
sunglasses. Never mind that the sun wasn't shining. I was ready for
a day on deck just as soon as I had my breakfast.

While the weather was rather flat, our spirits were high. It was Kerry's birthday so, after an exuberant round of 'Happy Birthday' and 'For She's a Jolly Good Fellow', we were all raring to get on deck and start our day of marine-mammal surveys.

It didn't take long to see why the Hebrides is such a special place for marine life. Heading out of the Sea of the Hebrides into the waters of the Little Minch, we encountered not the mythical Blue Men of the Minch but ten common dolphins. I was on my scheduled break at the time, so I had thirty free minutes to do as I pleased, which of course meant heading straight for the prow of the boat to look down at these graceful and agile creatures. Clearly, they were having a great time riding the bow wave.

It isn't really known why some marine mammals do this but, propelled forward by the pressure wave created by the moving vessel, they often don't even need to use their tail flukes to swim. They could be simply saving energy, but the individuals in this group of ten were constantly jostling each other to get the best position, and it looked like they were spending more than they were saving. It just looked like fun, and the amount of vocalisations picked up on the hydrophone certainly suggested that this was a social event. A little fun interlude from their day of travelling in search of food.

Watching the bow-riding dolphins was great, but the excitement of seeing them was quickly forgotten when I spotted not one but two adult male orcas. My first sighting of an orca in the wild did not disappoint. They were quite far away from us, but thankfully headed our way. Soon we would be able to see them closer up, to enjoy the splendid sight for longer than I might have hoped.

Even from a distance, their colossal dorsal fins stood straight and tall out of the water. It was impossible to estimate their size at that distance but, clearly, they were mature bulls with sprouted dorsal fins well over a metre if not closer to two metres in height. We knew we were looking at two male orcas, but we didn't yet know whether they were from the West Coast Community or perhaps from the Northern Isles population that frequent Shetland. As they swam closer and we could see them from the side it was clear that neither was the infamous John Coe as there was no telltale notch out of the bottom of their dorsal fins. Nor were either of them Floppy Fin as their dorsal fins were both upright.

With Moon already thought to be deceased, that left just Aquarius and Comet from the West Coast Community. Could it really be them?

Kerry checked the HWDT catalogue quickly and was pretty sure it was, but it wasn't until later that night we could confirm it from matching good ID shots of their dorsal fins we had just taken with the ID shots in the catalogue. We had chanced upon two of the possible remaining four males from the West Coast Community.

Unlike the common dolphins we had encountered earlier, the orcas were not particularly interested in us. They were headed our way only by coincidence and not because they wanted to come and play.

At one point they seemed to change course towards the boat. Tim cut the engine so as not to put them off or risk them getting too close to a turning propellor. While it is important to never approach a marine mammal and to give them sufficient distance, at least a hundred metres, sometimes the wildlife chooses to approach you. When it does, all you can do is soak it in and be grateful. It was a very fleeting close encounter as they changed course again and moved away from us.

Ensuring we left sufficient distance, we stayed with them for about an hour as Aquarius and Comet went about their day. We were no longer in our positions or looking out for other marine mammals or sea birds. Counting gannets could wait! We all stood on the port side of the boat with our cameras and binoculars, soaking up every second of this truly incredible encounter.

We all seemed to alternate between bursts of excitement and moments of quiet reverie. I had tried to capture decent photos of them, but after many attempts, I realised I was never going to cut it as a wildlife photographer. Kerry was well equipped with a big camera and lens and was getting much better photos. So, I decided to forget about looking at them through the lens and instead just take it all in with mental snapshots. It's all too easy to forget to enjoy these precious moments in life when we try to capture them on film. I am glad that I took the opportunity to just watch, learn and absorb their splendour. I still remember my feeling of utter disbelief that we had chanced upon this pair from the West Coast Community. It was amazing to share the experience with Tim and Tom, and we were all abuzz with excitement. It is hard to portray just how special a moment that was, a once-in-a-life-time experience.

As I watched the orcas go about their day, paying no interest other than their initial fleeting pass by the boat, I was really struck with a feeling of our insignificance on this planet, that we are just another species along with many others. We might be at the top of the food chain, but then in the marine world, so is the orca. It is our marine counterpart: equally wonderful, intelligent, caring, resourceful and powerful.

Having read so much about the Northern Resident orca popu-lation off Vancouver Island that are fish eaters, very sociable and inquisitive and eager to interact with passers-by, I'd had no qualms getting up close and personal with them in a kayak in Johnstone Strait. However, there was something different about Aquarius and Comet, whose lack of interest in the boat, and something about their speed and sharp turns in the water, made me feel that it was best to stay aboard *Silurian*.

Being predators of other marine mammals including minke whales, always hunting in groups of two or three, they need to stay silent and discreet. They are smart animals, smarter than we ever thought, and I know they wouldn't confuse a human with a porpoise, but their power and presence made me reluctant to mess with them. They were majestic though, commanding, elegant and captivating. A healthy respect for a wild animal is no bad thing.

We had seen them just past Neist Point Lighthouse, where they are frequently seen from land. Set on the most westerly point on the Isle of Skye, in the Glendale area, the lighthouse is located two kilometres from the car park, at the end of a path that is treacherous in bad weather. But on a nice summer's day it is the perfect place to go to with a picnic and look across the Little Minch to the Outer Hebrides.

In June and July, minke whales and basking sharks make almost daily appearances, seals and seabirds are both a 'given', but a sighting of the mighty orca would complete a truly lucky and spectacular day for any observer.

It was a magnificent day in the Little Minch and possibly the ultimate birthday present for a marine-mammal enthusiast like Kerry. As the day came to a close, we all celebrated with cake, tea and talk of our encounter with the UK's most extraordinary creature.

THE SHIANT ISLANDS AND SHEDLOADS OF SEABIRDS

Thursday 4 September 2014

Journey: Isle of Harris to Stornoway, Isle of Lewis
(Distance: 54 miles)

THE SHIANT ISLANDS, or 'fairy islands', are a cluster of small isles formed of columnar basalt rock and are actually an extension of the Trotternish Peninsula on Skye. They lie between the Little Minch Sea and the Greater Minch and are a truly spectacular sight with an enchanted feel about them, hence the name.

We approached on a grey and misty day, so they revealed themselves slowly and bit by bit. It was only when we were up close that we could make out any detail, and it took me a while to spot the small mountain sheep that were prancing nimbly around the steep, jagged coastlines. Our unsteady walking around the *Silurian* seemed clumsy in comparison.

Sheep are not the only things that have walked on this land. For many years black rats (*Rattus rattus*) also lived here, thought to have arrived after a shipwreck. Black rats were common on old sailing ships. Without much to prey on them and a plentiful supply of seabirds to eat, their numbers quickly escalated.

A few months before we passed, an eradication effort was begun, and by March 2018 the islands were declared rat-free.

The Shiant group is made up of three main islands: Garbh Eilean, Eilean an Taighe and Eilean Mhuire, more simply known as Rough, House and Mary Islands, which are pretty well direct translations from the Gaelic.

The Rough and House islands are linked by a thin strip of land, an isthmus, and combine to some 350 acres. It is the smaller Mary Island that is the more fertile thanks to the layer of Jurassic mudstone that covers the basalt which, when weathered, leaves the soil more fertile than most places in the Western Isles.

This fertility and a nice, halfway location between the Inner and Outer Hebrides attract many species of seabirds in great numbers, by far the highest concentration we had seen so far. It is said that tens of thousands of puffins breed in the burrows on the steep cliffs, but what I most remember seeing is thousands of kitti-wakes, shags, skuas and guillemots. Too many to count.

You can have a holiday here if you want, as there is a simple bothy located on House Island that is free to use provided you contact the island's owner. How beautiful it must be to wake up there with not another soul around, just hundreds of thousands of seabirds and the odd sheep for company.

It isn't a great place to anchor for the night though, as the sea can be pretty rough, and on this grey day we needed somewhere more sheltered. Our destination for the night was Stornoway in the Outer Hebrides.

By this time, we had settled into our various roles, and the routines had become second nature. Our day was set out in thirty-minute time slots.

Two people were assigned to Visuals. Beginning on the star-board side at the bow, they would continuously scan the ocean for signs of life, for blows, fins, spray, etc. After thirty minutes they would move to the port side and do the same again, both times looking out from abeam (90 degrees) to the bow.

We were attached to the main sail by harness and clip, to prevent us being thrown off. Sometimes it was calm on Visuals. At other times, notably on the day going out to St Kilda, we were face first into spray, rain and wind. When the swell was large, the downward lurch as the bow dipped to meet the sea could be jolting and an hour seemed like a long stretch.

This was no time to daydream. It was vital to stay fully alert every second. In smooth calm waters it was fairly easy to detect the blow from a porpoise or minke whale, but on rough days, when the water was choppy, a bobbing seal's head could be hard to spot. We were constantly straining our eyes in these conditions. Only the people on Visuals were allowed to call out a sighting.

If someone on the boat saw a marine mammal in your zone and you didn't call it, you can guarantee they would tell you afterwards and you would feel pretty stupid. If you saw a marine mammal or a creel buoy (see Lulu's chapter for their significance), you had to shout out 'Sighting!', alerting the person on relay to shout 'sighting' to the person on computer who would log the information.

After an hour on Visuals, it was time to rest the eyes and enjoy a break, quite often going below deck with a cuppa, having made one for everyone else as well, to warm the hands and get out of the rain. In good weather though, it was the best part of the day to join the skipper at the helm, relax in the sun and either have a little kip or a chat.

I was allowed to steer on one of my breaks as we sailed up the Little Minch and was surprised by how much harder it was than it looked. In rough waters with strong currents, it is surprisingly difficult to stay on track and, of course, it is important to stay on the guideline of our transect which is used in the scientific analysis.

The job on Relay was as it sounds: to relay the information from the guys on Visuals, gals in our case, to the person on the computer. You have to be quick, clear and concise, often prompting the person on Visuals for any information that they might have forgotten.

Relay is also responsible for doing bird and boat surveys. Less information is needed here: a count of each bird species and type of boat seen in each thirty-minute period.

This sounds easy enough, but some of those very small outcrops and islands, like the Shiant Islands and, more especially, the St Kilda group, are home to enormous numbers of birds. St Kilda is home to one of the world's largest colony of gannets, with over sixty thousand breeding pairs. Good luck counting all of those and, of course, guess who was on Relay as we left St Kilda. Yep, me.

I arrived at a number by splitting the sky into imaginary segments and doing a rough estimate of how many birds were in that segment, repeating the count three times to get an average for that segment. I then multiplied that by the number of segments. It was a valiant effort, I thought, but I suspect that a more real assessment would have been shedloads – too many to count. Sadly, there wasn't an option for this on the computer.

Computer amounted to listening on the hydrophone that had been placed in the water first thing each morning when the survey began. If there were no sightings, a pretty easy thirty minutes followed, listening to the sounds of the ocean. Armed with a clipboard from your Relay session, you would input the data on bird and boat counts, plus information such as sea state, height of swell, visibility and weather conditions, all the while listening for the telltale whistles and clicks of marine mammals.

As soon as you heard something you would press the record button to capture the sounds, but you had to keep the knowledge to yourself. Alerting those on Visuals that marine mammals were close would bias the data.

Of course, on Computer you also had to be quick and enter all the data being relayed from those on Visuals.

For those who suffered from seasickness, this was the worst rotation. A combination of not being able to see the horizon and the fumes from the engine, stronger below deck without the wind to diffuse them, for some meant emerging from below slightly green in the face.

Luckily, I have never been seasick, but one day when the swell was large and we were being jostled around, I did feel an uneasiness in my stomach. On Visuals, you can pre-empt a big wave and brace yourself but, below deck with no windows and just the computer screen to stare at, you are not prepared for either each wave of the sea or each wave of nausea.

While I wasn't sick, I was glad when my time was up and I could get some fresh air. There is nothing like a cold Scottish offshore breeze to wake you up and make you feel better.

The rest of our voyage that day was rather uneventful, a minke whale giving us a brief sighting before disappearing, and a couple of bobbing seals as we made our way into Stornoway, the capital of Lewis and Harris. The largest town in the Hebrides felt positively cosmopolitan after several nights with no civilisation in sight. It was the furthest north we would travel on our trip and the weather was glorious. As we slowly motored into the harbour the water was calm and the sky a deep blue.

Once we were moored, we took the opportunity to stretch our legs on land with a wander around the town. Some of the girls headed for the shops to stock up on pastries and chocolate, while I walked along the harbour admiring the array of boats, from beaten old skiffs to luxury yachts and the RNLI lifeboats. There was even what looked like a classic pirate ship, but it didn't look like it had been on any high seas for quite some time.

One of the highlights of our overnight stay was the simple pleasure of using the showering facilities at the harbour. Fresh water on the boat is limited and therefore so is showering. A brief, three minutes every few days in a cramped space was all we were allowed, so being able to take a nice long, hot shower with room to actually wash my hair and move around was something of a treat.

Feeling fresh and revitalised after my shower, I found a quiet place to sit in private and turned my mobile on to call Adam. Being newly engaged, four days with no contact felt like a long time and it was great to hear his voice and tell him that we had seen two of the West Coast Community. There was so much to tell but I had to get back to help out with dinner, so I rushed through the stories as quickly as I could. It must have been an incomprehensible monologue.

Weather permitting, next day we would make our way into the wider Atlantic for the islands of St Kilda, likely with an overnight stop. We would be without any reception for several days so I was thankful to have had this chance to call home. Suddenly, Stornoway and the Outer Hebrides felt a very long way from home.

STORNOWAY AND THE PLAYFUL MINKE WHALE

Friday 5 September 2014

Journey: Stornoway, Isle of Lewis to Lochmaddy, North Uist (Distance: 54 miles)

OUR ANCHORAGE FOR THE NIGHT in Stornoway harbour on the Isle of Lewis felt like a return to civilisation.

The harbour is beautifully sheltered and has a lovely outlook to Lews Castle, which is surrounded by lush woodlands. Seeing trees is unusual in the outer islands, and across much of Highland Scotland, as they are characteristically bare and vegetation sparse. It wasn't always this way. Much of the land was forested, but mankind cut it down over many centuries. In the days of the Vikings, trees were felled to make ships and houses and for fuel. During and after the Clearances, forests were cut down to make room for sheep farming, an industry which brought in more income than rent from tenant smallholdings known as crofts. Since then, more forests have been cleared for deer and grouse hunting. A changing climate over the centuries and the harsh and persistent winds that come in from the Atlantic made it hard for trees to recolonise, with peat bogs, heather and machair dominating much of the landscape instead. To make matters worse for the forests, continual grazing, particularly by sheep and deer, make it difficult for new saplings to develop.

Today, managing the deer population goes hand in hand with forest regeneration and the re-wilding movement. Typical images of the Scottish red deer capture this splendid creature with his large antlers, standing tall and proud on a barren hill with mountains in the background. The red deer is most at home in a forest environment but, ironically, their number prevents the forest from recovering.

For most of our trip, trees had barely been seen and so motoring into Stornoway and seeing the forests surrounding Lews Castle was such a treat. While the oceans have always brought me much peace and a sense of wonder, more and more I find solace in forests and woodlands. There is something magical about an ancient forest, so ethereal. When I walk through an old forest I experience a sense of healing, stillness and being alone, yet connected to our shared past, to people and to nature.

It was a shame that the forest was on the other side of the harbour and that we didn't have the time to go over and explore. We had just the one night at Stornoway, and the time was filled with a shower, dinner, bed and then in the morning, preparations for the day ahead. Exploring this little gem would have to wait for another trip.

As we left Stornoway, the sun was shining, the skies were blue and, being nicely rested and clean, we all felt invigorated. It seemed like a good time to practise our distance-estimating skills. With every sighting, whether marine mammal or creel buoy, whoever was on Visuals had to give an estimate of how far from the boat it was. This is harder than it sounds because, at sea, there are no reference points.

Whenever possible we would practise together, comparing our estimates against the read-out from a handheld laser distance recorder. This was not my strong suit. I might have been quick to spot marine life in the water, but did not find it easy to guess the distance.

First up was Kirsty with the laser distance recorder. She pointed it at the main sail furled up on a boat moored up in the harbour.

We took our guesses: 12 metres, 15 metres, 17 metres, 19 metres and me at 25 metres. Sara guessed the closest to the 17.8 metres from the handheld read-out, a good thing really as she had plans to become a marine-mammal observer and a good sense of distance at sea would serve her well. She took the laser distance recorded to select the next target and again we all shouted out our estimate. This time, it was a lobster pot sat on the side of the dock further away, with guesses of 25 metres, 27 metres, 32 metres, 40 metres and a wildcard 50 metres. And on we practised. Estimating distance is easier in the harbour, once there are stationary objects to measure and the distances are verifiable, not to mention easier with each subsequent guess where you have a reference point from the last guessed object.

We all improved the more we practised that morning, but it is quite another thing at sea when you spot the dorsal fin of a harbour porpoise some way off. Was it 50 metres, 100 metres, 200 metres? It is quite challenging, especially as the target disappears quickly, resurfaces in a different spot and can quickly change direction! In all honesty, it was something I was pretty terrible at for the majority of the trip, only really getting to grips with it towards the end.

While that does bring into question the accuracy of some of the data collected by a team of volunteers with little training in these things, I don't believe it detracts from the value of the data collected in terms of which animals were sighted and where. Sure, we might question the accuracy of the exact distance from the boats transects, but there is a rough pinpoint as to where each animal was spotted thanks to having an accurate reading of where the boat was.

With some satisfaction though, we noted that we had become more accurate as the voyage progressed.

Heading south, with the long Isle of Lewis and Harris on our right and the Shiant Islands on our left, our spirits were lifted again, this time by the difference in the weather. The day before had been overcast and grey, the longer views obscured by sea mist. Today, the sun was out, and we could see the Shiant Islands from a great distance.

Before, they had seemed mystical and romantic as they materialised in the mist. Now, seen in all their glory, they seemed almost tropical. It felt as if we were approaching a group of islands in Thailand. We were happily agreeing on this when a minke whale appeared.

By this point in the trip we had encountered quite a few minke whales but mostly the experiences were brief, just a single sighting before they disappeared. Being able to hold their breath for twenty minutes or so means they can vacate the area after a single, brief sighting. We would see a blow, but usually nothing else as they are not prone to aerial acrobatics like humpback whales.

The next time they come up they can be a great distance from the boat and lost to us visually. Many a time you won't even see the blow, you will simply smell their fishy breath. Their elusive behaviour and the fact that their breath smells like fish earned them the nickname 'Stinky Minke' among marine-mammal enthusiasts and scientists. I thought that, perhaps, this was a nickname given to them by the HWDT, but when I visited British Columbia a few years later, scientists at the various charities I was affiliated with also referred to them as Stinky Minkes. It appears that this is just an unfortunate universal epithet.

Today we were in luck though. A juvenile male decided it wanted to play. He swam from port to starboard and back again, rolling over and over, showing us his big white belly. We were delighted and excited by his display and apparently playful nature. It is really quite rare for this usually elusive species to interact so much and even Kerry, Tim and Tom, who spend so much time at sea, were quite enamoured. We cut the engine so as not to risk hurting him with the propellor, and so that Tim the skipper could leave his position and come and watch. We all had great fun following the whale from one side of the boat to the other for as long as he wanted to keep up the game.

It made me think of my cat back home, who loves to lie on his back, exposing his belly as an invitation for a belly rub. It is thought by animal behaviourists that it is a sign of greeting and great trust to expose one's belly like this. It leaves the cat in a vulnerable position and so they really must trust you to do this.

I wondered if the same applied to our minke friend. I like to think that he knew we were trustworthy and posed no threat. But equally, I hope he doesn't expose himself to every boat in this way. Not all boats and their passengers would delight in this behaviour as much as we did.

Whales, dolphins and porpoises are split into two main groups: the odontocetes, toothed whales such as dolphins and including the orca; and the mysticetes, who have baleen plates for filterfeeding instead of teeth. The rorqual whales, such as the minke, humpback and sei whales are part of the mysticete group. The minke is the smallest of the rorquals and easily distinguished from its relatives by a white band on each of its pectoral fins. These 'arm bands' were clearly evident on our playful friend.

As with the orcas that are seen in the Hebrides, sightings of minke whales also get reported, and it is clear from years of photographic identifications that there are some individuals that are seen annually. Using the photos we had taken and comparing them to the HWDT catalogue of known minke whales, it was evident that this individual wasn't known to the HWDT as a regular. However, there is a rather infamous minke whale in the Hebrides called Knobble, who has been sighted for thirteen years running, returning every summer to feed in the plankton-rich waters between the Inner and Outer Hebrides. While Knobble might be a resident whale in the summer months, nothing is known about where he or she goes in the winter. Kelsey is another frequently sighted minke whale in the summer months, of whom we know little.

It has become apparent that these waters, particularly the Sea of the Hebrides, a little to the south of where we encountered our playmate, are important to minke whales. In 2011, the HWDT, along with other charities like the Whale and Dolphin Conservation, proposed that the Sea of the Hebrides become a Marine Protected Area (MPA) to help safeguard these important and majestic animals.

In 2014 as I traversed these waters, it was still nothing but a pipedream, but continued data collection from volunteers on the *Silurian* and many other citizen science sightings all helped to strengthen the case. In 2020, after nine years of lobbying, this area became a protected area for minke whales just like the one we encountered. Success comes to those who persevere.

LEVERBURGH, ISLE OF HARRIS: GETTING INTO THE ROUTINE

Saturday 6 September 2014

Journey: Lochmaddy, North Uist to Leverburgh, Isle of Harris
(Distance: 54 miles)

WE ANCHORED THAT NIGHT in the harbour of Lochmaddy on North
Uist, a ferry port where the CalMac Ferries frequently dock. They
do kind of spoil the view for waterborne visitors but, of course, the
normal business life of the islanders must take priority. The landscape
is pretty flat and there are a few houses and a hotel but not a lot else.
The wind was picking up and when it started to rain we took shelter
around the table in the saloon with mugs of hot tea.

Since the weather forecast did not look great, Tim declared that
tomorrow would be 'pirate day'. We were not too sure what that
entailed when we were tasked with creating a pirate outfit that
night with strict instructions that we must be in full fancy dress
for breakfast. We spent our evening creating our outfits out of our
belongings and whatever we could find on the boat. It's amazing
what can be created out of an old cereal box and some tape, such
as Kate's parrot masterpiece that sat on her shoulder.

At 6am, we were awakened abruptly by the theme song from
Pirates of the Caribbean blasting from the saloon; another one of
Tim's pranks. We scrambled out of our beds and donned our
costumes, all slightly giddy with sleepiness but full of amusement
and excitement for the day ahead.

We were one crew member missing. Kate was nowhere to be seen. We joked that the obvious explanation was that she had been captured by pirates but, an hour later, we were starting to worry when we saw her walking down the road towards the harbour. She had decided to take an early walk and to breakfast at the Lochmaddy Hotel. She must have been up at the crack of dawn, if not before, and she was exceptionally quiet to sneak out of our room without my noticing.

With all crew now accounted for, costumes on and the music blaring, we were eager to start the day, although the sun had disappeared and the wind was strong.

It was too windy to travel far out to sea, so we remained in the relatively safe waters of the Little Minch. However, the wind meant one good thing: full sailing, no engine.

The requirements of constant speed and staying on the transect course meant that we motored most of the time. Not today though. The wind on this day allowed us to open the sails and enjoy the full force of nature.

With Lochmaddy behind us and the Little Minch in front, we had quite possibly our strangest encounter yet. As if written in a script, out of the misty grey sky appeared a real-life pirate boat. Okay, not actually a pirate ship but a real full-rigged ship with three masts. Now, this *was* a sailing ship, and it could not have timed its arrival better.

The first thing we noticed, aside from the sails, was the red and white flag at the helm. This was the Polish flag. The ship was called *Mir* and it turns out this is the world's tallest and fastest sailing ship. Upon some research after the trip, I learnt that the ship was built in 1987 in Poland, but she was now owned by the Admiral Makarov State Maritime Academy in St Petersburg and is a Russian training ship. At over 100 metres in length, this white and blue sailing boat dwarfed our 19-metre *Silurian*.

This trip was full of the unexpected and unbelievable. Considering it was rather bleak weatherwise, it was a really great day. First a pirate ship, then a harbour porpoise, seven common dolphins who enjoyed a quick bow ride and even one sneaky minke whale doing the usual trick of surfacing once, just enough to identify it, before taking a deep breath and then disappearing out of range before its next breath.

To top it all off, as we approached our anchorage at Leverburgh on the southern end of the Isle of Harris, a golden eagle flew overhead. A beautiful sight to behold. With a wingspan of over two metres, these are impressive birds of prey and we were incredibly lucky to see one as they are quite rare. While we were only a short distance north of Lochmaddy on North Uist, which is relatively flat, Leverburgh on the Isle of Harris is very mountainous, and a prime habitat for the golden eagle with its taste for mountain hares.

Leverburgh hasn't always been its name. Historically it was Obbe, meaning 'a bay' in Norse. Obbe was renamed in 1920 after its new owner, William Lever, the 1st Viscount Leverhulme, who planned to base a fishing empire here. With easy access to both the Minch and the Atlantic and their plentiful fish stocks, combined with lots of safe harbours and a sheltered bay, it seemed to be a perfect location. In 1925 he died of pneumonia, and his assets, including the village and fishing facilities, were sold for a fraction of their value.

While our anchorage was sheltered, it was not particularly spectacular or beautiful although Harris is famous for its beautiful beaches, especially those along the west coast road. Luskentyre Beach, with golden sand and clear waters of the most calming blues and greens, is simply stunning. Even better, likely you will have it to yourself should you visit, or maybe share it with a solitary dog walker.

Across the water from Luskentyre Beach is the Isle of Taransay, made famous by the TV programme *Castaway 2000*. This reality TV programme featured thirty-six men, women and children who were voluntarily marooned there for a year in an attempt to build a thriving community. As a reality TV programme it wasn't the most successful, but it did start the TV career of nature-loving, adventure-seeking, Ben Fogle.

While this part of the Outer Hebrides looks beautiful, it affords next to no protection from the strong winds and storms that sweep in from the Atlantic, making life on the island tough. You need to be hardy to live here, and the tough old Vikings did indeed make it their home in AD 900 during their invasion of Scotland. Of course, if anyone could handle it, it would be the Vikings.

The island has changed ownership and tenants many times but in 1974 it was left uninhabited and stayed this way, except for a few sheep, until *Castaway 2000*.

Ben Fogle loved his time on the island so much that he not only returned for his honeymoon but also tried to buy it when it went up for sale in 2011. Acquired for £11,000 in 1967, the going price by then was £2.2 million. Sadly for Mr Fogle, he was outbid, which strikes me as a shame as he had plans to establish it as a nature reserve.

All this being said, we didn't visit Taransay, or the nice beaches near Leverburgh, but stayed on the *Silurian* and got on with a boat life that had by now become routine, with meals and other food breaks established as markers of passing time. Perhaps because we got so cold standing at the bow for so many hours, or because the work was so repetitive, our minds never strayed far from our tummies.

Breakfast was at 6.30am, usually cereal and toast and a cup of tea while we readied ourselves and the boat for the day ahead. We would usually drop the hydrophone into the water and set off by motor around 7.30am. While we were a sailboat, unless we had conditions such as we had on pirate day, motor was the most reliable and consistent way to travel for scientific surveys.

Second breakfast was at 10am, and always something that could be eaten at the bow while holding on with one hand, such as a bacon sandwich or a fried-egg sandwich. Lunch would be a thick soup in a cup, again to be consumed while standing and working.

Even that was a bit ambitious on some of the rough days when we might send the contents flying overboard or worse still, over ourselves. We were never really off duty from lifting the anchor first thing until it was dropped again for the night, so everything had to be easy to eat.

Half-hour breaks every few hours were given over to quick trips to the toilet that could never be all *that* quick with so many layers to shed, and then to the tea round. We did usually have time to sit down on deck with our tea when the weather was nice, or the saloon if not. Afternoon snack was a cake, usually when the anchor was dropped around 5pm.

At the same time, we sorted out all the equipment and tidied the boat from the day's mess while someone else cooked. At dinner we would sit together at the table in lively spirits, renewing our acquaintance after another day of new experiences.

Tom, the first mate, was pretty good in the kitchen, thank goodness, and Kerry was a great baker. For our Leverburgh dinner she cooked up a Moroccan feast any Alawite would be proud of. No wonder that I put on several pounds on the trip, despite all my shivering.

ST KILDA, THE EDGE
OF THE WORLD

Sunday 7 September 2014

Journey: Leverburgh, Isle of Harris to Hirta, St Kilda
(Distance: 57 miles)

THE WEATHER PLAYS A MASSIVE PART in everyone's lives in the
Hebrides, residents and visitors alike. Before ours, many trips on
the *Silurian* had been unable to even reach the outer isles: the
wind and swell had been just too extreme. Even fewer have gone
beyond, to make the long and exposed journey to St Kilda, but we
were lucky. The weather stayed on our side . . . just.

St Kilda is an archipelago, with Hirta the largest island. Others
are Dun, Soay and Boreray. Anchored at Leverburgh, we readied
ourselves and the boat for the long journey to what felt like the
edge of the world, as it must have felt in the days when people
thought the world was flat. I confess that I knew nothing about St
Kilda before the trip and therefore had no expectations of what
awaited us, but I could tell from the team that even getting there
was going to be quite the journey. As it turned out, aside from
seeing the two orcas, Aquarius and Comet, the visit to St Kilda
was the highlight of the voyage for me.

While the main islands of the Outer Hebrides were formed
some 525 million years ago, St Kilda, created by a volcano, is
comparatively young at just 60 million years. Still, 60 million years
is a long time and the igneous rock that St Kilda is made of has

been eroded and shaped by relentless wind and waves, creating the jagged rocky islands with sheer cliff faces that we see today.

The remotest place in the British Isles, it lies forty-one kilometres from the most westerly point on the Outer Hebrides. Rough seas prevent travel most of the year and we were, in fact, the last boat to visit that year. After September, weather conditions make it just too risky. From St Kilda, the next landmass to the west is Canada but there is a lot of Atlantic Ocean to traverse first.

It took us all day to make the crossing, by far the roughest of our voyage. Clipping on when on deck, and especially at the bow when doing observations, was essential, and walking around the boat proved to be a good test of our sea legs.

We endured driving rain all day, but the skipper knew what he was doing, cutting the engine at just the right point on each large swell to ensure we didn't come back down with too much of a jolt. We were more exhilarated than frightened.

We had no sightings the whole way, with no other boats and no creel buoys to record. In that sense it was a bit quiet. There was likely plenty of marine life around, but the white spray of the waves and the swell made it hard to see dorsal fins and blows.

As the archipelago came into view, Tim said he could see a dragon. We all assumed this was another one of his jokes but, sure enough, as we got closer, there it was … a dragon. Well, actually it was the island of Boreray but from our approach it looked like a dragon lying down with his chin resting on the surface of the ocean, horns on the top of his head and the bumpy sloping mountain top forming the ridges of his spine curving down to his tail. Covered in grass, he was the classic green dragon and, with mist in the sky, it looked as if he had steam flaring from his nostrils.

We were rewarded as we approached Village Bay on Hirta Island, when a large pod of common dolphins escorted us in, bow riding and echolocating. They won't see many boats out this far, and were possibly as intrigued and excited as we were. They certainly made our approach even more magical and memorable.

Dropping anchor in the safety of the bay, we hopped onto the inflatable RIB at the back of the boat and motored to the concrete jetty that jutted out from the land. The little RIB was ideal for a quick dash and drop, but not ideal when raining as it was completely open to the elements.

Today, military staff live on the island all year on a rotation basis, as it is run by the Ministry of Defence (MoD) for range missile testing. Researchers also live here temporarily, as do officials from the National Trust for Scotland. There was just one lone man living there when we visited, an MoD employee who happily came out of his office to greet us and let us know where everything was. Then he left us to get on with exploring, clearly not one for too much idle chit-chat.

Formerly inhabited by about 180 islanders who lived in a settlement, called Main Street, in Village Bay, Hirta is now bereft of permanent residents, aside from those mentioned above. Their stone homes and crofts remain though, tended by volunteers organised by the National Trust for Scotland, which owns the islands, but it has been more than ninety years since they were evacuated in August 1930.

Back then, the land was owned by Sir Reginald MacLeod of MacLeod, who lived on the Isle of Skye. He was the 27th chief of the clan of MacLeod and laird of the islands. The St Kildans paid their rent to him by way of feathers and oils collected from seabirds. When years were particularly harsh, their rent might be reduced and food from the mainland gifted to them.

This relationship was not always easy or in favour of the islanders and by early 1930, after a hard winter and the deaths of two of their young women, they wrote a letter to the Secretary of State for Scotland, requesting that all villagers be evacuated and resettled into a life on the mainland. By this point, only thirty-six men, women and children remained, their already tough existence so much harder with fewer hands to hunt and tend to the sheep. While life was tough, it must have been a very difficult decision to make, as they were leaving behind the only life they had ever known. Never having left the islands they simply could not know what to expect of their new lives, except from fleeting glimpses based on what missionaries and tourists had told them.

The islands of St Kilda had been continuously inhabited for some two thousand years. That voyage on 29 August 1930 marked not only the end of human habitation but also the end of a way of life. I can only imagine how conflicted they must have felt as the boat pulled out of Village Bay. What a poignant movement it must have been. However, a snippet from the newspaper *The Bulletin*, archived in the National Records of Scotland, declared that there were no tears and even happy, excited faces, as their boat pulled away and they headed for their new life.

It was utterly fascinating to explore the abandoned village. The houses formed a line down 'Main Street', the only street on the island. They were dark, dirty and damp, and I could not imagine living in such conditions with my family, especially through the long winter months. They must have had strong minds as well as hardy bodies not to go insane. They must also have had a great community spirit.

A small church was built when missionaries were sent to conform their way of living to a more 'socially acceptable' standard with religion as its basis. Church was required of every islander, every day, if not multiple times a day.

The school room was off to the side of the church, a small wooden cladded room, the lower half painted brown, the upper half in white. On the walls was a map of the world, a map of Canada and a small bookshelf with a couple of books on it.

On one side of the room, there was a long wooden desk to sit six on a long bench, a little blackboard placed on the desk in front of each seat and three white quills stood upright in the wells. The ink and paper long gone.

On the other side of the room, no more than a metre away, was the teacher's lectern and a fireplace. With a window on two walls, it was well lit and when the fire was going it would be a cosy room, probably one of the nicest spots on the islands. Warm, dry and bright.

I imagine the kids loved coming to class although records from the missionaries and teachers remarked that they had little interest in learning to read or write, and that teaching even basic hygiene was a fruitless task.

On one of the walls there was a little shrine to their life with displays of their work and photos of the St Kildans. Some men wore bushy beards, with patched-up trousers and shirts, all of which looked rather grubby. Most wore a flat cap. Their clothing didn't look nearly suitable enough for the cold and wet weather that must have dominated their lives. Nor did they look to be the epitome of fitness and strength.

The women wrapped themselves in shawls, some up over their hair, others resting over their shoulders, and all appeared to part their hair in the middle with a low bun or ponytail. All wore a pair of gloves, mostly thick and warm, likely made from the wool from their Soay sheep. They typically dressed in long, dark, thick skirts which, I am sure, helped keep them warm, but didn't seem all that practical for their way of life.

In some of the photos they were dressed in their best clothes, alongside smart and statuesque but also rather pompous-looking missionaries and nurses who carried out their stints on the islands before heading back to mainland when they could no longer bear the conditions on St Kilda.

Some of the children in the photos wore shoes, but others went barefoot. In typical Victorian style, they all looked incredibly bored and sullen. I found it difficult to get a sense of whether they enjoyed their lives, but a cheeky suppressed grin on the face of one of the little boys gave me hope that they had fun like other kids.

I am sure their lives were mostly hard work; just getting food for dinner was a risky business. Although surrounded by the sea, fish was not the main item on the menu as fishing was often too dangerous. The sea could be extremely rough and wild, the location of the shoals unknowable, and simply getting off and on the island a treacherous task indeed.

Instead, they looked to the sky for their nutrition, to the fulmars and other sea birds that are so numerous here. Eggs formed a huge part of their diet. To reach the eggs which the fulmars lay in nests on the cliffs, men and boys would make the perilous climb down the rock faces, barefoot, to gather them. Many lost their lives.

Everything went to good use. The bird flesh and eggs were eaten, the feathers used to make pillows and bedding as well as pay the rent. Skin was used to make shoes, and oil from the birds' stomachs was burnt to keep warm. Fittingly, they have been nick-named 'bird people' or 'bird-men' in, for instance, the book *Land of Bird-Men: History of St Kilda* by Roberto Zanolla.

It was a truly special experience walking around the village and imagining what life must have been like. Spectacular scenery every way you look, a close-knit community to share your life with and the sense of a simpler way of living. Many of the things I yearn for in my own life and, I believe, many people around the world also seek. I came away feeling inspired by the tenacity and courage they must have had to endure life in St Kilda. It might be beautiful, but I am sure life there could also be brutal, especially in the winter months.

I felt galvanised to incorporate their sense of cooperation, hard work and community spirit into my daily life, starting with the rest of my trip on the *Silurian* with my teammates but first, a shower in the MoD facilities then back to the boat and my comfy bed. Now that is how to live life well in St Kilda.

SOAY, ST KILDA AND THE BLUEFIN TUNA

Monday 8 September 2014

Journey: Hirta, St Kilda to Leverburgh, Isle of Harris
(Distance: 61 miles)

ALTHOUGH WE COULD NOT DELAY our long return trip to the main group of Outer Hebridean Islands, it would have been a serious miss not to motor at least once around this magical cluster of islands.

The two sea stacks, Stac Lee and Stac an Armin, rise spectacularly out of the sea to a height of nearly two hundred metres. The tallest sea cliffs in the whole of Great Britain, they are a sight to behold, and it is pretty staggering to think that on Stac Lee there is a small bothy big enough for two men to collect sea birds and eggs. How they managed to get there is unreal.

The St Kildans used a small boat to reach the base of the stacks, weather permitting, and from there they would climb up the steep cliff face in just their socks, or even barefoot! A bothy to stay in was essential as it could be days, weeks or even months before they had a weather window sufficient to get back to Hirta. There are stories of a group of men being stranded there a whole winter.

When we approached St Kilda the day before, mist and grey skies had surrounded the islands and somewhat masked the full glory of the place. Today though, with the sun out and a cloudless sky, we could see their majesty in all its splendour. They might with good reason be called the 'islands on the edge' but, with so many birds around, it felt as if life was present in abundance. These two stacks alone are home to forty-five thousand breeding pairs of gannets.

It is this abundance in seabird colonies that led to St Kilda being listed as a UNESCO World Heritage Site (WHS) for nature, just as the history and importance of the islanders led this place to be listed as a UNESCO WHS for culture. It is one of only a handful of places around the world listed as a double WHS. Here I was, circumnavigating a double WHS of similar status to Machu Picchu in Peru. A lot closer to home but I imagine still as exciting and impressive.

As we passed around the northern side of Soay, we saw an astonishing sight. A common dolphin leapt from the water. That bit was fun but not peculiar, as we had seen many common dolphins by this time. What happened next was amazing though. A giant Atlantic bluefin tuna leapt into the air. The dolphin leapt again, apparently in pursuit.

It was possibly the last thing we were expecting to see, something we associate with the Mediterranean, not the cold Scottish waters off the Hebrides. Bluefin tuna are rare these days, especially this far north. Atlantic bluefin tuna used to frequent these waters, but a decline in herring numbers and an increase in fishing for large tuna meant a reduction in numbers until about 1990, when it was thought that there were none left in UK waters.

They are hard to mistake if you do see them: classic fish shape, they are torpedoes of power, up to three metres in length and weighing as much as 1,500 pounds. They can live up to an astonishing forty years.

It was just that one leap we witnessed so, to be honest, we all doubted our own eyes. Even though we all thought exactly the same thing, we collectively thought we must have been wrong and couldn't possibly have seen a giant tuna being chased by a dolphin. It wasn't until I got round to writing this book that I researched the matter. Indeed, Atlantic bluefin tuna are making a comeback in UK waters. They are special in that they are one of only a couple of fish that are warm-blooded, meaning they can control their body temperature.

One possible reason we are seeing more of them now is because, with climate change, the waters in Scotland are warming, which could be increasing the number of mackerel and herring that they feed on. Or perhaps their stocks are just recovering as fishing pressure has been removed. Whatever the reason, it is rather awesome that we are getting these giant fish back in our waters.

As you would imagine for a fish of such importance and interest, they are being monitored by Marine Scotland and the Marine Management Organisation (MMO). The MMO grant licences to specific skippers to take part in a catch-and-release tagging programme so that the numbers and location of tuna can be ascertained.

The stocks are protected and at present, no one, recreational or commercial, is allowed to fish for these bluefin tuna. But no doubt it won't be long before they are targeted by fisheries. After all, it's hard to resist a fish that is worth tens of thousands of pounds.

St Kilda certainly delivered us our portion of non-mammal wildlife, but it is also a place that frequently delivers on the marine-mammal front as well. In particular, it is a good place for orca sightings. I later spoke to Angus Campbell, the owner and skipper of Kilda Cruises based on the Isle of Harris, who is also one of the trained skippers for the tuna-tagging programme. Angus takes groups of nature enthusiasts and thrillseekers out to St Kilda regularly and has been doing so for years.

We had a great chat one evening, although it was tricky to catch every word through a crackling phone connection and his unfamiliar-to-me Hebridean accent. Angus sees orcas most years. He has even seen a group of eight swimming together on the eastern side of Boreray. What a sight that must have been, with the most spectacular backdrop. He also witnessed a group hunting seals and tossing them into the air and has even seen several calves over the years. This is wonderfully affirming, but they are sadly not our West Coast Community. They are likely the Northern Isles population that occasionally venture into west Scottish waters in the summer months.

The journey back from St Kilda to Leverburgh was uneventful with very few sightings, just one bobbing head of a seal at a distance, too far away to determine which species and a solitary harbour porpoise as we neared Leverburgh. The glorious weather though made up for the lack of sightings. On our way to St Kilda we had battled against strong winds, large swell and rain that soaked us through but our return journey, just the next day, was the exact opposite. Such is the weather in Scotland.

We all enjoyed the sunshine and smooth ride. For the first time on this trip our skin was exposed to the elements, and it felt great. With our arms and faces catching some vitamin D, we all soaked in the warmth. It truly felt like a holiday now.

We had a real sense of team spirit, and friendships had been formed all round. While we all kept our eyes on the water, we enjoyed relaxing and telling stories from home. I sat with Kirsty whenever I could, helping her put together a more detailed plan for her Master's thesis. She planned to investigate whether there were yet any detectable impacts of climate change on the marine-mammal population in the Hebrides. No doubt our bluefin tuna sighting would feature. Both my parents were teachers before they retired and, with this experience, I was really starting to see the appeal of teaching. It was a great feeling helping another develop and I felt sure that I would keep in touch with Kirsty afterwards as she wrote up her thesis.

We were quite an eclectic group from different backgrounds and at different life stages, but there were never any tensions within the group. Everyone pitched in with chores and kept spirits light and fun. That night, we played a few games of Bananagrams, a silly word game which had become a familiar nightly routine.

My cold finally disappeared and I felt that beautiful combination of being at once full of life and quietly at peace, a feeling that only comes to me at sea, or somewhere remote where I can check out from everyday life.

I was relaxed from not having access to my phone, emails or social media, as a media detox is always a good idea, but I was happy to get back to Leverburgh and enough of a signal to let Adam know I was back from the Edge of the World.

LOCH SHIELDAIG AND THE SUPERPOD OF COMMON DOLPHINS

Tuesday 9 September 2014

Journey: Leverburgh, Isle of Harris to Loch Shieldaig
(North-west coast of Scotland mainland)
(Distance: 45 miles)

THE FINE WEATHER OF THE DAY before had sadly left us, forcing a return into thermals and waterproofs. It was also time to say goodbye to the Outer Hebrides and begin our return journey across the Minch. We all felt that we had been extraordinarily lucky with orca and other sightings, and that it could not continue. It would surely be a quiet journey home with nothing special to report, we thought. The Hebrides, however, had other plans.

Out of nowhere, five common dolphins made a beeline for the port bow of the boat. Seven more followed, joining on the starboard side but swiftly moving forward to tussle for a hitch on the bow waves. Another group arrived, then another, then another and, suddenly, we were surrounded.

For as far as we could see, on either side of the boat, common dolphins leapt and splashed, moving quickly but socially through the water. There must have been at least a hundred, the highest number of marine mammals I had ever seen in one place and by far the most spectacular sight.

We were also delighted to see several calves, which always makes my heart sing. Seeing several babies is a good sign that the population is doing well and, of course, there is nothing better than seeing a mother and her baby together, whether they be human or dolphin. The bond that mammalian mothers have with their offspring is a beautiful thing, and a thousand times more powerful and emotive to me since I became a mother myself.

The babies stay close at all times, often swimming in mum's slipstream. This keeps those bonds strong, and also means they are close to their food supply. The mother nurses her young for around six months before they start to become independent.

Just as quickly as they joined us, they departed. Play was over and it was time to move on with their day. They were all vocalising loudly in their excitement, and clearly all got the message pretty rapidly, and were on their way. It was quite incredible to see such an organised and coordinated change in behaviour.

We waved them goodbye and wished them well and were once again amazed by all the wildlife there is to see in this special place.

As we approached Loch Torridon, a ragged indent of mainland Scotland, we had a couple of sightings of harbour porpoises, which now seemed rather run-of-the-mill. It is all too easy to become complacent when it should always be a delight to see a marine mammal.

The scenery as we made our way down the fifteen miles of sea loch was so impressive and breathtaking it would have taken more than a porpoise to tear our attention away. Loch Torridon is surrounded by some of Scotland's most iconic mountains, namely Liathach, Beinn Alligin and Beinn Eighe, which are all over 3,000ft. Collectively they are called the Torridon Hills and are made of Torridonian sandstone. I could imagine how beautiful they must look when covered in winter snow, pristine except for a few tracks left by mountain hares and, in days gone by, the grey wolf.

We anchored for the night at Shieldaig, in Loch Shieldaig, which is within Loch Torridon, a picture-postcard village of whitewashed cottages with grey roofs and one of the prettiest sights we had seen. Or was that St Kilda, or the Isle of Rum or the Shiant Islands . . . You get the picture, there are just so many beautiful spots in the Hebrides.

Herring has been fished in these waters since Viking times, and it is still an important fishing town. In fact, the name Shieldaig is a Viking word meaning 'loch of the herring'.

While most of the surrounding area has long been cleared of trees, there is an island which is heavily forested and now a nature reserve. Little Shieldaig Island has been owned by the National Trust for Scotland since 1970, and its dense pines stand in contrast to the barren landscape of the mainland, which has few trees, only low-lying grasses.

It is believed that these pines were planted to give the fishermen somewhere to hang their nets at night. However, this species of pine is not thought to be native to this patch of land and there are plans to gradually replace them with their native counterpart.

While the day hadn't begun in beautiful conditions it had ended well, and Loch Shieldaig was the perfect place to drop anchor and watch another blazing Hebridean sunset.

A couple of the others got on the RIB and headed over to the local pub for a few drinks that night, and one of my few regrets on this trip was not taking the opportunity to go with them. In my mind it would have been a night to remember: drinking whisky, telling old legendary stories and ceilidh dancing with an old, bearded man who also played the fiddle into the wee small hours.

The reality though, was a tame night in a quiet local with just a handful of patrons having a quiet drink after work. I'm not sure why I didn't go ashore. It's strange, but I felt unsure about leaving our little boat and maybe having a few too many to drink and paying for it the next day on rough waters.

Instead, I woke up early, clear-headed, relaxed and virtuous, and very much enjoyed my morning cup of coffee on deck while the others had their breakfast in the saloon. The air was fresh and crisp and the water calm. It was the perfect way to start the day, looking out to sea, surrounded by mountains and nature at their best. The Hebrides had won a special place in my heart and I wish I could have bottled the feeling of serenity I felt at that moment.

KYLE RHEA AND
THE FULL MOON

Wednesday 10 September 2014

Journey: Loch Shieldaig to Loch Scresort, Isle of Rum
(Distance: 60 miles)

WE WERE WELL AND TRULY on our way home now, making the
return journey to the Isle of Rum today and then tomorrow
continuing to Tobermory. I woke reflecting on the experiences I
had so far enjoyed on this trip, how it had exceeded my expectations and I would go home a happy person even if we didn't see
another marine mammal. I was also looking forward to getting
back to my fiancé, Adam. I was eager to tell him about the trip,
especially seeing two of the West Coast Community.

We were on a tight schedule today as the journey would take
us through the narrow passage of Kyle Rhea that sits between the
Isle of Skye and mainland Scotland. You can only get down this
narrow stretch of fast-flowing water on the correct tides. At points,
the strait is just six hundred metres wide, and tides as fast as ten
knots have been recorded. It was important to get moving and to
get the timing just right.

While it might have been a brisk, cold start to the day, it
certainly warmed up. We motored underneath the Skye bridge,
which connects the island to the mainland, and shed layers of
clothing to enjoy the warmth of the sun on our arms and faces. It
was a picturesque setting with beautiful houses dotted along the

shoreline and mountain scenery on both sides. Fishing is evidently a big industry around these parts, which is not surprising considering the strong tidal currents that continuously bring in nutrient-rich waters which support a great deal of marine life. It felt like every five seconds we were calling out 'Creel buoy!', but by this time we were a well-oiled machine and had our distance estimates pretty accurate.

We didn't have to wait long to get some proper sightings. Today's treat was harbour porpoises. We had been so spoilt for marine-mammal sightings on this trip that to see a harbour porpoise had become not all that exciting, but today was different. Usually, you just see one or two when approaching land as they rarely venture out far into open water, but they are usually pretty docile and in small numbers. As they come out of the water just sufficiently to breathe, a dorsal fin and a small blow might be visible, but they are shy and never approach a boat.

Today though, pod after pod appeared in a short space of time. At breakfast, Kerry had predicted we might see some strange behaviour today. She had noticed on her trips that after a full moon harbour porpoises tend to be bolder, more gregarious, displaying aerial acrobatics more suited to common dolphins. She was right, they were definitely swimming faster and more playfully, with several individuals leaping from the water.

The strait opened as we approached the southerly tip of the Isle of Skye, where the water was like a mirror. The sun was out and there was no wind, so I climbed in a state of bliss, or was that full of nerves, to the crow's nest to watch for sightings. It was breathtaking to absorb the beauty of the Hebrides from this vantage, a different perspective that only deepened my reflective mood. In a way this was surprising as I am not one for heights. Going rock climbing with Adam led to an embarrassing situation when I froze at the top of a practice wall, tiring myself by clinging on for dear life. In the crow's nest though, it didn't feel high or scary. It felt like being at the top of the world and totally at peace.

Approaching the Isle of Rum, we were treated to yet more playful harbour porpoises. It really had been a day of the porpoise and I was glad to have rekindled my joy at seeing this elusive and mysterious creature. This was our second time in the safe harbour of Loch Scresort and, as we dropped anchor, we all felt in high spirits. As a group we had bonded well. You get to know people surprisingly well when living together at close quarters and having to do everything as a team.

This was the summer of 2014, and there was a craze going round for people to throw a bucket of ice-cold water over their heads to raise awareness and encourage donations for the ALS charity for those suffering from motor neurone disease. I had mentioned earlier in the trip that a friend had nominated me, and my shipmates kindly reminded me that I had not yet completed my challenge. On our last night, with an abundance of cold water all around us, I suggested that rather than pour a bucket of ice over my head, I jump into the sea. The others needed no encouragement and before I knew it, we were all in our swimming costumes lined up along the port side of the boat.

Tom sat in the crow's nest and filmed my obligatory speech about raising money for ALS. Freezing in just my swimsuit, I spoke quickly, eager to get it over with, but also rather excitedly as, although it was cold, we were all having so much fun. At the very moment I was nominating my crewmates, Karen, Tim and Tom to do the challenge next, I was caught off guard when Tim threw a bucket of seawater over my head. I shrieked like a little girl but there was nothing else to do other than jump right in.

The water was so cold it took my breath away. I could just about muster enough air to encourage the others to join me, which one by one they did. It was so cold, I couldn't feel my legs after a couple of minutes but, I have to say, I hadn't laughed so much in a long time. One by one, we clambered back onto the boat in the most unladylike fashion, given that we couldn't feel our limbs, grabbing our towels to dry off and get dressed.

Over full mugs of steaming hot tea, we warmed up and laughed about the fun we had had that day and through the whole trip. I knew I had made real friends on this trip and, in fact, still keep in touch with several of the girls years later.

Karen accepted the challenge and duly jumped in with me, but it didn't go unnoticed that Tim and Tom stayed dry and warm. Reminiscing about this story made me wonder why I didn't catch them off guard and push them in or throw a bucket of cold Hebridean Sea water over them, but alas, I didn't. Perhaps I knew better than to trick a trickster!

Tomorrow, we would make our final journey back from the Isle of Rum to the Isle of Mull where our trip around the Hebrides would come to an end. The only question that remained was what delights we would witness on our last day. There was still time for another sighting of the West Coast Community. I remained hopeful as always and settled down for a peaceful night's sleep.

FROM RUM TO OBAN:
LEAVING THE HEBRIDES

Thursday 11 September 2014

Journey: Loch Scresort, Isle of Rum to Tobermory, Isle of Mull
(Distance: 50 miles)

ON OUR LAST DAY WE ALL WOKE in high spirits, and many a conversation revolved around the incredible luck we had shared.

We all kept our eyes firmly on the water, hopeful that we would get our last spectacular sighting of the trip, possibly a minke whale as we once again passed through the 'minke triangle' of the Small Isles. Of course, I was hopeful for one last sighting of the elusive and captivating West Coast Community, but this was not to be.

As we pulled into Tobermory, it seemed our luck had run out, but we knew better than to grumble. We had seen a tremendous amount of wildlife in only ten days, even making it as far as St Kilda, had learnt new skills, made new friends and, most importantly, had contributed to science and to the conservation of marine life in the Hebrides.

It might not have been the type of holiday that most people sign up for, but it is one that will enrich their lives. Despite it being a working holiday, I came away feeling rested, restored and more balanced than I had in years.

In ten days of surveying we had travelled over 515 miles of the Hebridean seas, seen some 420 cetaceans (whales, dolphins and porpoises), had fifty-eight seal sightings, shedloads of seabirds, several white-tailed and golden eagles, gravity-defying cliffhanger sheep and even one leaping bluefin tuna.

The cetacean sightings, of course, included a superpod of common dolphins bow riding with their calves, full-moon leaping porpoises, very friendly minke whales and of course, the *pièce de résistance*, the mighty orca. Not one, but two of the West Coast Community, Aquarius and Comet. The Hebrides certainly is a special place for wildlife watching.

Just because we were not 'on effort' any more, and keeping our eyes peeled for that telltale blow or dorsal fin, our work was not done. We now had the job of cleaning the boat thoroughly, inside and out, to ready her for the next group. We all pitched in and, remaining in high spirits, the job was done quickly and rather enjoyably, leaving time at the end for some group photos and general tomfoolery.

Taking advantage of being back in civilisation, we all headed for the shower block to freshen up before making our way to dinner in one of the best restaurants in the Scotland. 'Café Fish' looks rather unassuming from the outside, located on the upper floor of the CalMac pier at the end of the main street. The decor is coastal, as you would expect, with a very simple and unfussy feel about it, plain wooden tables and chairs, not a tablecloth in sight and a very affordable menu. It certainly doesn't scream 'best restaurant in Scotland', even though the views over the harbour are spectacular. However, when the food began to arrive we understood why it has its outstanding reputation. Kerry and Tom, who live in Tobermory, were greeted with warm embraces and waves across the restaurants from friends. Clearly, Café Fish is as popular with the locals as it is with the tourists.

The freshly baked bread was delicious and the seafood was so fresh, simply cooked and utterly delectable, all beautifully accompanied with a chilled glass of Sauvignon blanc. On our way back to the boat, we decided it would be rude not to pop in for one more with the locals and so we all enjoyed another drink (or two!) at the Mishnish pub.

The Mishnish dates back to 1869 and is jam-packed with memorabilia from shipwrecks in the local region. It is the bright yellow building in between the bold red and blue buildings. You really can't miss it! It proved to be a real locals' pub and has a friendly atmosphere where everyone knows everyone else. It was surprisingly busy for a Thursday night. I am not sure if it was the several glasses of wine that were consumed or the lingering rocking you experience when you come back onto land after time at sea, but when I walked back to the boat that last night, I definitely felt a telltale sway.

The next morning, after our final breakfast in the saloon, we all boarded the CalMac ferry for our return to Oban. We headed straight for the top deck and got our binoculars out, and it was as if we all were on duty again, resuming our stations. We didn't have to wait long until our first sighting of a common dolphin and, so ingrained was it now, all called 'Sighting!' and began to list all the important data (that no one was recording). A crowd gathered and we took pleasure in showing the tourists where to look and telling them all about the amazing marine life in the Hebrides.

Before I knew it, I was back on the Scottish mainland with just a train ride to Glasgow and a short flight back to Heathrow and the congestion of the M4 and . . . just like that, I was home and the Hebrides seemed a million miles away.

EPILOGUE

Looking Back

AQUARIUS AND COMET WERE THE FIRST wild orca I ever saw, and my encounter with them is among my most precious memories. That it happened in their breathtaking home environment of the Hebrides was a blessing, a beautiful backdrop in which to watch in complete awe. Knowing that I am one of the last people to ever see Comet brings me much sadness, as does knowing that someone else will have the last sighting of Aquarius and John Coe.

We have learnt that the West Coast Community is a small, isolated family that does not mix with any of the other orcas that visit the waters around Scotland. In the last 30 years, no calves have been seen and the ten members of this group are now likely down to just two remaining elderly males. Their exclusive diet of other marine mammals separates them from visiting orcas, completing an isolation that, in turn, has led them to be highly inbred. This is one factor in their catastrophic reduction in population size.

Another key factor is the devastating impact of PCBs from the legacy chemicals that contaminate our waters. Thankfully, they are now banned, but it appears that the West Coast Community included some of the most polluted marine mammals on the planet. Responsible disposal of remaining PCBs is crucial if we are to return our waters to good health.

Additional stresses, such as entanglement in fishing gear, plastic pollution, fish farms, underwater noise from ADDs, shipping noise, and use of military sonar only add to their plethora of pressures and stress. Climate change, if it has not already, will certainly become another threat not only to orcas but to all the oceans' inhabitants. That our marine apex predator is flailing is a clear wake-up call. We are not living in harmony with our oceans and must strive to find a better balance.

My hope is that this book will have informed, inspired and encouraged you to learn more about the incredible wildlife we are blessed to share our planet with, and to be aware of how we are impacting it. I have purposely referred to our West Coast Community by their names rather than catalogue numbers as I strongly believe that, by seeing these animals as living, feeling beings that are intelligent and capable of love, compassion and empathy, we might better relate to them and to the crisis in our oceans.

Orcas have family bonds as strong as our own. They collectively care for their young, old and sick, ensuring that they are fed and never left behind. Culture is key to their survival, with knowledge and wisdom passed from one generation to the next, the elders leading and teaching the young. Their days are spent cooperatively working and playing together, remaining close, with physical contact maintaining their bonds and affirming their love.

The more we learn about them, the more we realise that we are not alone with our high-order thinking and emotional intellect. Orcas, among many other species, match or even surpass us. I find them to be humbling and inspiring. There is so much we can learn from nature if we leave our ego at the door and embrace the lessons that we so desperately need to hear.

LOOKING FORWARD

Ways you can help

Volunteering on the Silurian with the Hebridean Whale and Dolphin Trust

One of the most rewarding things you can do, both for conservation and for yourself, is to volunteer for the *Silurian*, as I did. Surveying with the HWDT directly contributes to a decades-long dataset which is now of international importance. For ten days you will experience life as a marine-mammal surveyor, witnessing first-hand the bountiful seas around the staggeringly beautiful Hebridean Islands. It is the perfect way to explore the area, learning through experience. You can find out more about volunteering with the Hebridean Whale and Dolphin Trust on board the *Silurian* here: https://hwdt.org/silurian

If you can't commit your time to joining one of these surveys you might want to become a member of the HWDT or sponsor an animal or make a donation to the charity. You can even sponsor the West Coast Community and get annual updates on the remaining members. All these things help provide vital funds to aid the HWDT to continue their research and education outreach. https://hwdt.org/join-support-2

Report your sightings

If you have been lucky enough to spot a marine mammal (especially an orca), whether from land, ferry or boat, don't forget to call it into the HWDT. By doing this you become part of their citizen-science database, which is crucial to our understanding of what is going on with marine life in the Hebrides. Don't forget to let them know where you were, what date and time, what species you saw and how confident you are, any behaviour seen, how many were present, including any calves and general weather conditions and sea state. Don't worry if you can't remember all that, just report what you do know and ideally grab a photo to upload as a lot of information can be gleaned in this way. Report your sightings to: https://whaletrack.hwdt.org/report-sighting. Or make life even easier and download the Whale Track app, which is the quickest and easiest way to report sightings. As we know, the West Coast Community don't stay in the Hebrides so, if you spot one of the remaining orcas while you are in Ireland, you can report your sighting to the Irish Whale and Dolphin Group (https://records.iwdg.ie/sighting.php) or to the Sea Watch Foundation (https://www.seawatchfoundation.org.uk/sightings-form/) if you are elsewhere in the British Isles.

More citizen-science projects are run by the Whale and Dolphin Conservation (WDC) charity. Shore watch relies on trained volunteers to conduct ten-minute marine-mammal surveys at eighteen specific locations along the Scottish coastline. Every volunteer that has been trained uses the same equipment, the same reporting sheets, the same time frame. This is the best way to get an accurate idea of what animals are seen (or not seen) at specific locations. These surveys are carried out year-round to gain a better idea of seasonal distribution. Shorewatch has eleven sites that cover the complete range where the West Coast

Community is likely to venture. These landbased observations offer an impact-free way of watching and monitoring marine mammals around Scotland. You can find out more about how to sign up here: https://www.wdcs.org/national_regions/Scotland/shorewatch/index.php

Dead strandings

If you encounter a dead stranded marine mammal when you are in Scotland, contact the Scottish Marine Animal Stranding Scheme (SMASS) to report it. Gather as much information as possible, such as the species or a description of the animal, when and where you found it, its length and the general condition of the remains. Photographs are always helpful so, if you have your phone or camera handy, take a few shots of the entire animal (preferably with something next to it to provide a scale), a close-up of the head, genitals and any visible cuts and marks. This will enable the team at SMASS to make initial observations on the species, age, gender and possible cause of death. You can text (or phone) all this information to: 07979 245893 or email it to: reports strandings.org. As you have seen in this book, a dead marine mammal can provide vital information by way of a necropsy if the body hasn't decomposed too much. Even if a necropsy can't be performed, data on where and when a stranding has occurred and for which species can still tell scientists a lot about the health of the population. For more information on reporting a dead marine mammal take a look at the SMASS website: http://www.strand-ings.org.

Live strandings (BDMLR)

If the animal is still alive, time is of the essence. The people to call in this instance are the British Divers Marine Life Rescue (BDMLR). With trained marine medics and specialist gear situated throughout the UK, they are ready 24 hours a day, 365 days a year to respond. The hotline to call is 01825 765546. You should note much the same information as a dead stranding, where and when you found the animal, what species it is or a description of it, the number of animals, any obvious signs of problems. Do not be tempted to approach the animal, just call in the information and await further guidance. Marine mammals in distress on a beach can behave in an unpredictable manner.

You could of course go one better, and volunteer to become a marine medic. You don't need to be a diver, just be happy to stand in the cold and wet, or heat of the day, for a long time, be enthusiastic and passionate about marine life, and they will train you with everything else you need to know. The BDMLR has been running for over thirty years and has rescued countless animals. In 2018 alone, they received 1,458 callouts for marine animals in distress, which is almost four every single day. As a registered charity they are always in need of both people volunteering their time to help but also financial support through donations. For more information on the BDMLR visit their website: http://www. bdmlr.org.uk.

Whale-watching

The time for viewing marine mammals in captivity has passed. We must now move towards seeing these amazing creatures where they belong, in the wild. I promise you that you will have a far better experience and learn more by seeing them in their natural habitat than you will from them performing tricks in a show. That being said, we must be responsible in the wild by following common-sense guidelines: by keeping our distance from wildlife, giving animals plenty of space and not following or harassing them in any way, slowing down if in a boat and leaving them alone if they display signs of distress. There are whale-watching tours wherever in the world there are whales, and some places have stricter guidelines than others. Go only with reputable companies that take the welfare and conservation of the marine mammals seriously. It pays to do a bit of homework before you sign up to anything. Too many boats around the same group of cetaceans spoil the experience for you and are stressful for the animals. Look out for those that have their WiSe accreditation for responsible marine wildlife viewing (https://www.wisescheme. org).

The Hebridean Whale Trail

If you don't have your sea legs and would prefer to look for whales from the shore, be sure to check The Hebridean Whale Trail (https://hwdt.org/the-hebridean-whale-trail). The HWDT have gathered all their data and experience to put this set of locations together to give you the best chance of viewing marine life, and seeing marine mammals from the shore is really rewarding. There is zero interference from you, so you come away knowing you have not impacted their day. Instead you can marvel, guilt-free, at their beauty. One of my favourite things to do is a good hike to a quiet viewing place with a picnic in my backpack, sun cream, glasses and binoculars at the ready. It is a joy to while away the hours, taking in the scenery and fresh air and then if you get lucky, your patience being rewarded with a sighting or two. Having a camera with a good lens is always a benefit so you can zoom in and get some good photos. Again, if you do spot a cetacean, don't forget to report it to the HWDT. Other places that have been reported as being great orca-viewing spots from land are Ardnamurchan Lighthouse (on the mainland, just north of the Isle of Mull), Stoer Head Lighthouse (on the west coast mainland) and Neist Point Lighthouse on the Isle of Skye.

Be a conscious consumer

Make a conscious effort to reduce the amount of single-use plastic you have in your life. Single-use plastics are among the most wasteful products man has created, serving us for only the shortest amount of time but taking hundreds, if not thousands, of years to break down. Many end up in landfill where the chemicals leach into the soil. Others are burnt, releasing toxic particles, and others end up in the ocean where marine animals either consume them, confusing them with food, or become entangled in them or otherwise injured. As the plastics slowly break down they separate into smaller and smaller amounts until finally they are invisible but still very much present as microplastics which are ingested into the food we eat. Sadly, single-use plastics are pervasive in our culture, from food packaging to the products on our bathroom shelves and cleaning products, nearly every purchase we make has been wrapped in single-use plastic. It takes a lot of effort to avoid it, but things are becoming easier, so I urge everyone, wherever possible, to find alternatives. Being intentional in our purchases is one of the greatest differences we can make as individuals. Feel empowered that you can make a difference.

Be vigilant with your own litter. Make sure it goes into the bin and not onto the ground. Make a habit of picking up litter you see when you are out and about. Take a bag with you when you go out for a walk and pick up anything you see on the way. All rivers lead to the sea so, even if you don't live anywhere near the coast, collecting litter is important. You could join or even set up your own 'urban beach clean' where you live. The 'NotWhaleFood' campaign run by WDC was for this exact purpose: urban beach cleans to prevent plastics and other rubbish getting into the ocean that marine mammals might mistake for food.

Safely dispose of PCBs

As we saw with Lulu, PCBs are possibly the biggest threat to the orca globally. In the UK, some 80 per cent of PCBs that were produced before they were banned in the 1970s still need to be responsibly destroyed. If you have equipment that you know contains PCBs then it is your responsibility to ensure that it is disposed of in a safe and legal manner. You can get more guidance here: https://www.gov.uk/guidance/polychlorinated-biphenyls-pcbs-registration-disposal-labelling#ban-on-pcbs

Light at the end of the tunnel

Despite the predicted collapse of 50 per cent of the world's orca population there are some populations that live in relatively pristine environments and could well double in size. Therefore, the total number of orcas might remain stable, but restricted to the last remaining clean areas. Populations that could do well are the Alaskan residents, Antarctica Type C, the Canadian Northern Residents in Telegraph Cove that I have been lucky enough to see twice, and those in Norway, the Crozet Archipelago and the Eastern Tropical Pacific. It is of great importance to protect these populations and to ensure that we maintain minimal human disruption to these populations.

The Northern Residents hold a very special place in my heart as it was reading about these whales in Erich Hoyt's book that first sparked my interest. Whale-watching generates ever-growing income in places such as Tofino and Norway. We must remain vigilant to keep recreational interest in these animals as non-invasive or disruptive as possible. Diving with orcas, as is promoted in Norway, is hugely controversial and can be incredibly disruptive to a pod of orcas while they are feeding. Do your research before you book a trip to make sure that whale-watching boats are staying within the guidelines and regulations.

I hope this book has highlighted how incredible orcas are and just how lucky we are to have them in our waters although, sadly, we will lose the West Coast Community. It is just a matter of time. We can only hope that as they pass on, their bodies are found so that we can learn all we can about them. Knowledge is power and if we can fully understand where they came from, how they became so isolated and why they were unable to thrive, hopefully we can apply this knowledge to other small populations.

We are having a profound impact on our marine life, even on the oceans' apex predator, and have a moral duty to safeguard the amazing creatures we are blessed to share this planet with. The quote at the start of this book is from Rachel Carson, author of the milestone book *Silent Spring*. She stated: 'One way to open your eyes to unnoticed beauty is to ask yourself, "What if I had never seen this before? What if I knew I would never see it again?"' Our chances of seeing any of the remaining West Coast Community are quickly running out, but let's not, one day, be saying that about all orcas, everywhere in the world.

Acknowledgements

THERE ARE MANY PEOPLE I NEED TO THANK for their help and support in writing this book. While I have an academic background, I am by no means a leading authority on the orca or a pioneering researcher, only a humble orcaholics and wildlife enthusiast. Credit must go to those people who have dedicated their lives to understanding more about these amazing animals and who work tirelessly to protect them. There are too many to name them all, but the following list includes those who have touched me most: Dr Michael Bigg, Dr John Ford, Dr Ingrid Visser, Eva Saulitis, Ken Balcombe and, of course, Erich Hoyt. It was Erich's book that first piqued my interest in orca research, so I thank him for not only his work on orcas but also for inspiring me and so many others. I am indebted to him for his support in writing this book and for so kindly writing the foreword.

I would like to thank Sir David Attenborough who has inspired, informed and motivated me for decades to care about our planet. A personal handwritten note from him was the encouragement I needed to finally reach out to a publisher to have this book become a reality.

I would like also to extend a heartfelt thank you to the following people and organisations who informed my writing, imparted knowledge and encouraged me along the way.

First and foremost, I am so grateful to the Hebridean Whale and Dolphin Trust for all the work they do to protect marine life around the Hebrides as well as all their education and outreach work. It was on one of their research surveys that I was so fortunate as to encounter my first wild orcas, Comet and Aquarius. I learnt so much on that trip and was inspired to write this book. Thanks to all my crewmates on the *Silurian*: Kate, Karen, Sara, Gemma, Kirsty, Tom, Tim and Kerry, for making it such a fun, interesting and rewarding two weeks. Playing Bananagrams without them just isn't the same. Thanks, in particular, to Sara Bisset for her enthusiasm for this book and reading early drafts.

It was great to have the knowledge and passion of HWDT's Dr Lauren Hartney-Mills on board with this project, with special appreciation for all the fact checking and knowledge-sharing. Sincere thanks to Rob Lott from Whale and Dolphin Conservation for fact-checking the final manuscript and his encouraging feedback and to Dr Filipa Samarra for checking wording on the Type 2 ecotypes.

Sincere thanks are also due to the following:

Dr Lance Barrett-Lennard and the Vancouver Aquarium Marine Mammal Research Team for taking me on as a summer intern to learn more about orcas, and for the research they carry out. It was an honour to listen to. and learn from them all, not to mention a joy spending hours analysing drone footage of the Northern Residents. I never tired of entering their world.

The Strawberry Isle Marine Research Society (SIMRS), in particular Jessica Edwards and the late Rod Palm. I felt incredibly privileged to share a cup of tea with Rod Palm in his galley and hear some of his fascinating orca stories from his incredible life on the water of Clayoquot Sound. SIMRS and Tofino have held a special place in my heart ever since and I applaud the work that this charity carries out in this unique place, as well as their involvement with the First Nations who respect and protect nature like no one else.

Dr Andy Foote, the leading authority on this group of orcas, took the time to talk to me over Skype and reply to emails and many questions. His knowledge was so helpful in pulling this book together. Without Andy's research, we would know scarcely anything about this group of orcas.

Dr Andrew Kitchener and Dr Georg Hantke from the National Museums Scotland for talking to me about their experience of retrieving the male orca from South Uist in 2015.

Angus Campbell from Kilda Cruises for taking the time to share his experience of seeing orcas around St Kilda.

Lari Don, a children's author based in Edinburgh, for talking to me about the myths and legends in Scottish folklore.

Daniel Brooks is a former Hebridean whale-watching tour guide and now West Coast Community enthusiast with great knowledge and passion. His comments on the draft manuscript were so helpful and his enthusiasm for the book was greatly appreciated.

The team at Sandstone Press, who originally published the hardback edition of this book, and especially my editor, Robert Davidson, for understanding the importance of the orca's story, for his attention to detail and for turning my words into something more eloquent and intelligible. I'd also like to thank Andrew Simmons and the team at Birlinn for their efforts and thorough edits for this paperback edition.

My dear friend Anni Ludhra-Gent, aka my accountability partner, kept me motivated to keep writing, gave feedback and made early edits of the book. Her constant support for this endeavour has been a godsend.

Jordan Szatkowski, a relative (second cousin once removed, I believe) and aspiring Texan marine biologist helped with research on marine-protected areas for the book.

Kay, for helping me overcome imposter syndrome and to believe in myself. I don't think I would have had the courage to resume work on the book without her guidance.

My parents and sisters, who have always supported and encouraged my dreams and interest in marine biology and allowed me to decorate my bedroom as a shrine to marine mammals.

I am eternally grateful to my husband, Adam, for supporting me in writing this book. Without his support and encouragement and belief in my writing, I don't think I would have finished. There are no words to say how thankful I am that he came out to British Columbia, not once but twice, so I could have my time with the whales, and especially for getting in a boat and a kayak to go whale-watching. The biggest gesture of his love for me.

Thank you to my two children, Quinn and Owen, for being my reason and motivation to write. I hope you grow up loving the wildlife we are privileged to share our planet with and that you have a life filled with adventures in nature.

Last, but by no means least, to our orcas of the Hebrides: John Coe, Nicola, Floppy, Comet, Moon, Lulu, Aquarius, Puffin, Occasus and Moneypenny. On behalf of mankind, I apologise that we failed you by not protecting and cherishing our oceans, your home, as we should have done. Thank you for simply being here, in all your magnificent splendour, for inspiring and educating us. I hope this book goes some way to helping other populations of your kind to live long, peaceful and healthy lives.

Bibliography

Introduction

Lott, R. 2018. Northern (High) Lights. *BBC Wildlife Magazine*, May 2018. 68–73.

May, P. 2013. *Hebrides*. London: Quercus Editions.

Katz, B. 2017. 'UK killer whale contained staggering levels of toxic chemicals'. *Smithsonian Magazine*. 4 May 2017. https://www.smithsonianmag.com/smart-news/body-uk-orca-contained-staggering-levels-toxic-chemical–180963135/

Desforges, J-P., Hall, A., McConnell, B., Rosing-Asvid, A., Barber, J.L., Brownlow, A., De Guise, S., Eulaers, I., Jepson, P.D., Letcher, R.J., Levin, M., Ross, P.S., Samarra, F., Vikingson, G., Sonne, C. and Dietz, R. 2018. 'Predicting global killer whale population collapse from PCB pollution'. *Science* 361 (6409), 1373–6. doi:10.1126/science.aatl953

Hebridean Whale and Dolphin Trust. 2018. Hebridean Marine Mammal Atlas. Part 1: *Silurian, 15 years of marine mammal monitoring in the Hebrides*. A Hebridean Whale and Dolphin Trust (HWDT) Report, Scotland.

Foote, A.D., Hooper, R., Alexander, A., Baird, R.W., Baker, C., Ballance, L., Barlow, J., Brownlow, A., Collins, T., Constantine, R., Dalla Rosa, L., Davison, N.J., Durban, J.W., Esteban, R., Excoffier, L., Fordyce Martin, S.L., Forney, K.A., Gerrodette, T., Gilbert, T., Guinet, C., Harrison, M.B., Li, S., Martin, M.D., Robertson, K.M., Samarra, F.I.P., de Stephanis, R., Tavares, S.B., Tixier, P., Totterdell, J.A., Wade, P., Wolf, J.B.W., Fan, G., Zhang, Y. and Morin, P.A. 2021. 'Runs of homozygosity in killer whale genomes provide a global record of demographic histories'. *Molecular Ecology* 30,6162–77, doi: 10.1111/mec.16137

JOHN COE: Residents, transients and everything in between

Hartny-Mills, L. 2021. 'Famous killer whales "John Coe" and "Aquarius" back in Hebridean seas after being spotted off the English coast for the first time'. Hebridean Whale and Dolphin Trust blog. 20 May 2021.

BBC News. 'Shark attack suspected on killer whale John Coe'. 28 January 2015.

https://www.bbc.co.uk/news/uk-scotland-highlands-islands-30980599

Barrett-Lennard, L.G., Matkin, C.O., Durban, J.W., Saulitis, E.L. and Ellifrit, D. 2011. 'Predation on gray whales and prolonged feeding on submerged carcasses by transient killer whales at Unimak Island, Alaska'. *Marine Ecology Progress Series* 421, 229–41. doi:10.3354/meps08906

Sea Watch Foundation. 2013. 'Scientists stunned as John Coe goes east!' 23 August 2013. http://www. seawatchfoundation.org.uk/scientists-stunned-as-john-coe-goes-east

BBC News. 'UK's only resident killer whales spotted in Cornwall'. 6 May 2021. https://www.bbc.co.uk/news/uk-england-cornwall-57011389

Webster, L. 2021. 'Killer whales John Coe and Aquarius back in Hebrides after Cornwall sighting'. *The National.* 20 May 2021. https://www.thenational. scot/news/19316605.killer-whales-john-coe-aquarius-back-hebrides-cornwall-sighting/

Garrard, P. 2021. 'Killer whale update: John Coe and pal are spotted in English Channel'. *Hebridean Whale and Dolphin Trust.* 14 June 2021. https://hwdt. org/news/john-coe-seen-in-english-channel

Pitman, R. and Durban, J. 2010. 'Killer whale predation on penguins in Antarctica'. *Polar Biology* 33, 1589–94. doi:10.1007/s00300-010-0853-5

Dwyer, S. and Visser, I. 2011. 'Cookie cutter shark (*Isistius Sp.*) bites on cetaceans, with particular reference to killer whales (*Orca*) (*Orcinus orca*)'. *Aquatic Mammals* 37, 111–38. doi:10.1578/am.37.2.2011.111.

Pitman, R.L., Deecke, V.B., Gabriele, C.M., Srinivasan, M., Black, N., Denkinger, J., Durban, J.W., Mathews, E.A., Matkin, D.R., Neilson, J.L., Schulman-Janiger, A., Shearwater, D., Stap, P. and Ternullo, R. 2016. 'Humpback whales interfering when mammal-eating killer whales attack other species: mobbing behaviour and interspecific altruism?' *Marine Mammal Science* 33 (1), 7–58. doi:10.1111/mms.12343

Durban, J.W., Fearnbach, H., Burrows, D.G., Ylitalo, G.M. and Pitman, R.L. 2017. 'Morphological and ecological evidence for two sympatric forms of Type B killer whale around the Antarctic Peninsula'. *Polar Biology* 40, 231–36. doi:10.1007/s00300-016-1942-x

Torres, L., Pinkterton, M.H., Pitman, R., Durban, J. and Eisert, R. 2013. 'To what extent do type C killer whales (*Orcinus orca*) feed on Antarctic toothfish (*Dissostichus mawsoni*) in the Ross Sea, Antarctica?' *Commission for the Conservation of Antarctic Marine Living Resources.* WG-EMM–13/29

Pitman, R., Durban, J., Greenfelder, M., Guinet, C., Jorgensen, M., Olson, P., Plana, J., Tixier, P. and Towers, J. 2010. 'Observations of a distinctive morphotype of killer whale (*Orcinus orca*), type D, from subantarctic waters'. *Polar Biology* 34, 303–06. doi:10.1007/s00300-010-0871-3

National Geographic. 2019. 'Mysterious new orca species likely identified'. 9 April 2019. https://www.nationalgeographic.co.uk/2019/03/mysterious-new-orca-species-likely-identified

Bigg, M.A., Olesiuk, P.F., Ellis, G.M., Ford, J.K.B. and Balcomb, K.C. 1990. 'Social organisation and genealogy of resident killer whales (*Orcinus orca*) in the coastal waters of British Columbia and Washington State'. International Whaling Commission Report of the Commission Special Issue, 12, 383–405.

Hoyt. E. 2019. *Orca: The Whale Called Killer.* 5th edn. Toronto & Buffalo, NY: Firefly Books.

Heise, K., Barrett-Lennard, L., Saulitis, E., Matkin, C. and Bain, D. 2003. 'Examining the evidence for killer whales predation on Steller sea lions in British Columbia and Alaska'. *Aquatic Mammals* 29 (3), 325–34.

Strawberry Isle Marine Research Society. http://strawberryisle. org

Bourton, J. 2010. 'Two killer whale types found in UK waters'. *BBC Earth News.* 5 January 2010. http://news.bbc.co.uk/earth/hi/earth_news/newsid_8440000/8440002.stm

Foote, A.D., Newton, J., Piertney, S., Willerslev, E. and Gilbert, T.M. 2009. 'Ecological, morphological and genetic divergence of sympatric North Atlantic killer whale populations'. *Molecular Ecology* 18, 5207–17. doi:10.1111/j.1365–294X.2009.04407.x

Smith, K. 2021. 'First killer whale match between Scotland and Norway'. *Scottish Field.* https://www.scottishfield.co.uk/outdoors/wildlifeandconservation/first-killer-whale-match-between-scotland-and-norway

Selbmann, A., Deecke, V.B., Fedutin, I.D., Filatova, O.A., Miller, P.J.O., Svavarsson, J. and Samarra, F.I.P. 2020. 'A comparison of northeast Atlantic killer whale (*Orcinus orca*) stereotyped call repertoires'. *Marine Mammal Science* 37(1), 268–89. https://onlinelibrary.wiley.com/doi/abs/10.llll/mms.12750

Moura, A.E., Kenny, J.G., Chaudhuri, R.R., Hughes, M.A., Reisinger, R.R., de Bruyn, P.J.B., Dahlheim, M.E., Hall, N. and Hoelzel, A.R. 2015. 'Phylogenomics of the killer whales indicates ecotype divergence in sympatry'. *Heredity*, 114 (1), 48–55.

NICOLA: Matriarchs and menopause

Wright, B.M., Stredulinsky, E.H., Ellis, G.M. and Ford, J.K.B. 2016. 'Kin-directed food sharing promotes lifetime natal philopatry of both sexes in a population of fish-eating killer whales, *Orcinus orca*'. *Animal Behaviour* 115, 81–95.

Barrett-Lennard, L. 2000. 'Population structure and mating patterns of killer whales (*Orcinus orca*) as revealed by DNA analysis'. PhD Thesis. University of British Columbia.

Baird, R.W. and Dill, L.M. 1996. 'Ecological and social determinants of group size in transient killer whales'. *Behavioural Ecology* 7, 408–16.

Centre for Whale Research, Friday Harbour, WA. https://www. whaleresearch. com

Hoyt, E. 2019. *Orca: The Whale Called Killer.* 5th Edn. Toronto & Buffalo, NY: Firefly Books.

Veirs, S., Veirs, V. and Wood, J.D. 2016. 'Ship noise extends to frequencies used for echolocation by endangered killer whales. *PeerJ.* 4, e1657. doi:10.7717/peerj.1657

Brent, L.J.N., Franks, D.W., Foster, E.A., Balcom, K.C., Cant, M.A. and Croft, D.R 2015. 'Ecological knowledge, leadership and the evolution of menopause in killer whales'. *Current Biology* 25 (6), 746–50. doi:10.1016/j.cub.2015.01.037

Photopoulou, T., Ferreira, I.M., Best, P.B., Kasuya, T. and Marsh, H. 2017. 'Evidence for a postreproductive phase in female false killer whales *Pseudorca crassidens*'. *Frontiers in Zoology*, 14, 30. doi:10.1186/sl2983–017–0208-y

Ellis, S., Franks, D.W., Nattrass, S., Currie, T.E., Cant, M.A., Giles, D., Balcomb, K.C. and Croft, D.P. 2018. 'Analyses of ovarian activity reveal repeated evolution of post-reproductive lifespans in toothed whales'. *Scientific Reports* 8, 12833.

Foster, E.A., Franks, D.W., Mazzi, S., Darden, S.K., Balcomb, K.C., Ford, J.K.B. and Croft, D.P. 2012. 'Adaptive prolonged post-reproductive lifespan in killer whales'. *Science* 337, 1313.

Nattrass, S., Croft, D.P., Ellis, S., Cant, M.A., Weiss, M.N., Wright, B.M., Stredulinsky, E., Doniol-Valcroze, T., Ford, J.K.B., Balcomb, K.C. and Franks, D.W. 2019. 'Postreproductive killer whale grandmothers improve the survival of their grandoffspring'. *Proceedings of the National Academy of Sciences of the United States of America* 116(52), 26669–73.

COMET: *The trailblazer*

Derry Journal. 2015. Dopey Dick's Derry visit on BBC.

Parfit, M. 2013. *The Lost Whale: The True Story of an Orca Named Luna.* New York: St Martin's Press.

The Helen Hamlyn Centre for Design and Art. 2016 projects: Our Future Foyle. https://www.rca.ac.uk/research-innovation/helen-hamlyn-centre/research-projects/2016-projects/our-future-foyle

Olesiuk, P.F., Ellis, G.M. and Ford, J.K.B. 2005. 'Life history and population dynamics of northern resident killer whales (*Orcinus orca*) in British Columbia'. Fisheries and Oceans Canada. Research Document, 2005/045.

ABC News. 2018. 'From the archives: Humphrey the whale leaps into hearts of Bay Area residents in 1985'.

Delta & Dawn: Sacramento river rescue. The Marine Mammal Center, 2007.

Taylor, M. 2018. 'Killer whales seen in River Clyde'. *The Guardian*. https://www.theguardian.com/environment/2018/apr/22/killer-whales-seen-in-river-clyde

Daily Motion video footage: https://www.dailymotion.com/video/x6i83i8

Horton, H. 2018. 'Killer whales "put on a show" in River Clyde'. *Telegraph*. 22 April 2018. https://www.telegraph.co.uk/news/2018/04/22/killer-whales-put-show-river-clyde/

MOON: *The circle of life*

Sperm whale explosion in Taiwan: https://www.express.co.uk/news/world/1078252/sperm-whale-exploded-in-street-taiwan

Hodgins, N. 2016. 'Why are beached whales taken to landfill?' Whale and Dolphin conservation. 31 March 2016. https://uk.whales.org/ 2016/03/31 /why-are-beached-whales-taken-to-landfill/

Smith, C.R. 2003. 'Bigger is better: The role of whales as detritus in marine ecosystems'. In *Whales, Whaling and Marine Ecosystems*, James Estes (ed). Berkeley, CA: University of California Press.

Scottish Marine Animal Stranding Scheme. http://www.strandings.org

Kitchener, A. 2015. 'The tricky process of collecting marine specimens'. National Museums Scotland blog. 14 April 2015. https://blog.nms.ac.uk/2015/04/14/the-tricky-process-of-collecting-marine-specimens

Kitchener, A. 2016. 'Collecting marine specimens: The killer whale skeleton'. National Museums Scotland blog. 13 February 2016. https://blog.nms.ac.uk/2016/02/13/collecting-marine-specimens-the-killer-whale-skeleton

LULU: *The world's most polluted whale*

BBC News. 2016. 'Killer whale found dead on Tiree identified as Lulu'. 6 January 2016. https://www.bbc.co.uk/news/uk-scotland-glasgow-west-35244417

Bawden, T. 2016. 'Lulu: whale's death may mean the end for Britain's orcas'. *Independent*. 8 January 2016. https://www.independent.co.uk/environment

/nature/lulu-whale-s-death-may-mean-the-end-for-britains-orcas-a6803011.html

Scotland Now. 2016. 'Killer whale Lulu found dead off Scottish island as extinction fears grow'. *Scotland Now*, 7 January 2016.

Scottish Marine Animal Stranding Scheme (SMASS) 2016 annual report.

Maclennan, E., Leaper, R., Brownlow, A., Calderan, S., Jarvis, D., Hartney-Mills, L. and Ryan, C. 2020. 'Estimates of humpback and minke whale entanglement in Scotland'. International Whaling Commission Report of the Commission. SC/68B/HIM/01.

Northridge, S., Cargill, A., Coram, A., Mandleberg, L., Calderan, S. and Reid, B. 2010. Entanglement of Minke Whales in Scottish Waters; An Investigation into Occurrence, Causes and Mitigation. Sea Mammal Research Unit University of St. Andrews – Final Report to Scottish Government.

Scottish Entanglement Alliance (SEA). https://www.scottishentan-glement.org/

Whale Entanglement Team (WET). Observes Male Orca Become Entangled in the Monterey Bay National Marine Sanctuary. https://marinelifestudies.org/4516-oo-entangled.html

Scullion, A.J., Harrop, H., Munro, K., Truluck, S. and Foote, A.D. 2021. Scottish killer whale, *Orcinus orca*. Photo Identification Catalogue, 2021.

Waugh, D.A., Suydam, R.S., Ortiz, J.D. and Thewissen, J.G.M. 2018. 'Validation of growth layer group (GLG) depositional rate using daily incremental growth lines in the dentin of beluga (*Delphinapterus leucas* (Pallas, 1776)) teeth'. PLOS One. doi:10.1371/journal.pone.0190498

Foote, A.D., Hooper, R., Alexander, A., Baird, R.W., Baker, C., Ballance, L., Barlow, J., Brownlow, A., Collins, T., Constantine, R., Dalla Rosa, L., Davison, N.J., Durban, J.W., Esteban, R., Excoffier, L., Fordyce-Martin, S.L., Forney, K.A., Gerrodette, T., Gilbert, T., Guinet, C., Harrison, M.B., Li, S., Martin, M.D., Robertson, K.M., Samarra, F.I.P., de Stephanis, R., Tavares, S.B., Tixier, P., Totterdell, J.A., Wade, P., Wolf, J.B.W., Fan, G., Zhang, Y. and Morin, P.A. 2021. 'Runs of homozygosity in killer whale genomes provide a global record of demographic histories'. *Molecular Ecology* 30, 6162–77, doi: 10.1111/mec.l6137

Ross, P.S., Ellis, G.M., Ikonomou, M.G., Barrett-Lennard, L.G. and Addison, R.F. 2000. 'High PCB concentrations in free-ranging Pacific killer whales, *Orcinus orca*: effects of age, sea and dietary preferences'. *Marine Pollution Bulletin* 40(6), 504–15.

Jepson, P.D., Deaville, R., Barber, J.L., Aguilar, A., Borrell, A., Murphy, S., Barry, J., Brownlow, A., Barnett, J., Berrow, S., Cunningham. A., Davison, N., Ten Doeschate, M., Esteban, R., Ferreira, M., Foote, A., Genov, T., Gimenez, J.,

Loveridge, J., Llavona, A., Martin, V., Maxwell, D.L., Papchlimitzou, A., Penrose, R., Perkins, M., Smith, B., de Stephanis, R., Tregenza, N., Verborgh, P., Fernansez, A. and Law, R. 2016. 'PCB pollution continues to impact populations of orcas and other dolphins in European waters'. *Nature Scientific Reports*, doi:10.1038/srepl8573

Morelle, R. 2017. '"Shocking" levels of PCB chemicals in UK killer whale Lulu'. BBC News, Science and Environment. 2 May 2017. https://www.bbc.co.uk/news/science-environment-39738582

Embury-Dennis, T. 2018. 'Bereaved mother orca finally drops calf after carrying corpse for unprecedented 17 days'. *Independent*. 12 August 2018. https://www.independent.co.uk/news/world/americas/orca-killer-whale-mother-tahlequa-dead-baby-southern-resident-washington-a8488486.html

FLOPPY FIN: Nature versus nurture

Bigg, M. 1982. 'An assessment of killer whale (*Orcinus orca*) stocks off Vancouver Island', International Whaling Commission Report of the Commission, 32, 655–66.

Visser, I. 1998. 'Prolific body scars and collapsing dorsal fins on killer whales (*Orcinus orca*) in New Zealand waters'. *Aquatic Mammals* 24 (2), 71–81.

Towers, J.R., Ellis, G.M. and Ford, J.K.B. 2015. 'Photo-identification Catalogue and Status of the Northern Resident Killer Whale population in 2014'. Canadian Technical Report of Fisheries and Aquatic Sciences 3139.

Jourdain, E. and Karoliussen, R. 2018. *Identification Catalogue of Norwegian Killer Whales: 2007–2018*. 3rd edn. Andenes, Norway: Norwegian Orca Survey.

Stenersen, J. and Similä, T. 2004. *Norwegian Killer Whales*. Tringa forlag.

Hof, P.R. and Van der Gucht, E. 2007. 'Structure of the cerebral cortex of the humpback whale, Megaptera novaeangliae (Cetacea, Mysticeti, Balaenopteridae)'. *Anatomical Record* 290, 1–31.

Oceana, Protecting the World's Oceans. 2013. 'Disabled killer whale survives with help from its pod'. 21 May 2013. https://usa.oceana.org/blog/disabled-killer-whale-survives-help-its-pod.

Wright, B.M., Stredulinsky, E.H., Ellis, G.M. and Ford, J.K.B. 2016. 'Kin-directed food sharing promotes lifetime natal philopatry of both sexes in a population of fish-eating killer whales, *Orcinus orca*'. *Animal Behaviour* 115, 81–95.

Cowperthwaite, G. 2013. *Blackfish*. Magnolia Pictures.

Hargrove, J. 2015. *Beneath the Surface: Killer Whales, SeaWorld and the Truth beyond Blackfish*. London: Palgrave Macmillan.

You, L. 2019. 'The activists fighting to free China's captive killer whales'. 20 August 2019. *Sixth Tone*. https://www.sixthtone.com/news/1004434/the-activists-fighting-to-free-chinas-captive-killer-whales.

Daly, N. 2019. 'Release of whales from notorious Russia "whale jail" complete'. *National Geographic*. 10 November 2019. https://www.nationalgeographic.com/animals/article/russia-moves-orcas-and-belugas-from-whale-jail

Baizhong, G., Au, B. and Coroneo-Seaman, J. 2021. 'Inside China's booming ocean theme parks'. 19 February 2021. China Dialogue Ocean. https://chinadialogueocean.net/16296-inside-chinas-booming-ocean-theme-parks

PUFFIN: Hebridean hubbub

Baird, R.W. and Dill, L.M. 1996. 'Ecological and social determinants of group size in transient killer whales'. *Behavioural Ecology* 7, 408–16.

Barrett-Lennard, L.G., Ford, J.K.B. and Heise, K.A. 1996. 'The mixed blessing of echolocation: differences in sonar use by fish-eating and mammal-eating killer whales'. *Animal Behaviour* 51, 553–65.

Ford, J.K.B., Ellis, G.M., Matkin, D.R., Balcomb, K.C., Briggs, D. and Morton, A.B. 2005. 'Killer whale attacks on minke whales: prey capture and anti-predator tactics'. *Marine Mammal Science* 21, 603–18.

Ford, J.K.B. and Ellis, G.M. 2005. 'Prey selection and food sharing by fish-eating "resident" killer whales (*Orcinus orca*) in British Columbia'. Research Document, 2005/41. Fisheries and Oceans Canada.

Wright, B.M., Stredulinsky, E.H., Ellis, G.M. and Ford, J.K.B. 2016. 'Kin-directed food sharing promotes lifetime natal philopatry of both sexes in a population of fish-eating killer whales, *Orcinus orca*'. *Animal Behaviour* 115, 81–95.

Deecke, V.B., Nykanen, M., Foote, A. and Janik, V.M. 2011. 'Vocal behaviour and feeding ecology of killer whales *Orcinus orca* around Shetland, UK'. *Aquatic Biology* 13, 79–88. doi:10.3354/ab00353

Riddington, G., Radford, A. and Gibson, H. 2020. 'The economic contribution of open cage aquaculture to Scotland: A review of the available economic evidence'. Salmon and Trout Conservation Scotland Sustainable Inshore Fisheries Trust. January 2020.

Watson, J. 2021. 'Animal welfare campaigners sound out ban on acoustic seal deterrent'. *The Sunday Times*, 21 February 2021. https://www.thetimes.co.uk/article/animal-welfare-campaigners-sound-out-ban-on-acoustic-seal-deterrent-x5fbl7pm7

Southall, B.L, Bowles, A., Ellison, W., Finneran, J., Gentry, R., Greene, C.Jr., Kastak, D., Ketten, D., Miller, J., Nachtigall, P., Richardson, W., Thomas, J. and Tyack P. 2008. 'Marine mammal noise exposure criteria: Initial scientific recommendations'. *Aquatic Mammals* 33, 411–521.

Edwards, R. 2021. 'Salmon companies rapped for breaking rules on shooting seals'. *The Ferret.* 25 February 2021. https://theferret. scot/salmon-companies-rapped-shooting-seals

Findlay, C.R., Ripple, H.D., Coomber, F., Froud, K., Harries, O., van Geel, N.C.F., Calderan, S.V., Benjamins, S., Risch, D. and Wilson, B. 2018. 'Mapping widespread and increasing underwater noise pollution from acoustic deterrent devices'. *Marine Pollution Bulletin* 135, 1042–50. doi: 10.1016/j.marpolbul.2018.08.042

Morton, A.B. and Symonds, H.K. 2002. 'Displacement of *Orcinus orca* (L.) by huge amplitude sound in British Columbia, Canada'. *ICES Journal of Marine Science* 59, 71–80.

Todd, V.L.G., Williamson, L.D., Jiang, J., Cox, S.E., Todd, I.B. and Ruffert, M. 2021. 'Prediction of marine mammal auditory-impact risk from Acoustic Deterrent Devices used in Scottish aquaculture'. *Marine Pollution Bulletin* 165.

FIDRA. 'Is salmon farming costing the Earth?' Best Fishes Project. https://www.bestfishes.org.uk/scottish-salmon-farming-impacts

Scottish Environment LINK. November 2020. Acoustic Deterrent Device Statement. https://www.scotlink.org/wp-content/uploads/2021/03/LINK-ADD-Statement-March–2021-Final-2.pdf

Royal Navy. Exercise Joint Warrior. https://www.royalnavy.mod. uk/news-and-latest-activity/operations/united-kingdom/exercise-joint-warrior

Piantadosi, C. and Thalmann, E. 2004. 'Whales, sonar and decompression sickness'. *Nature* 428, 1–2. doi:10.1038/nature02527a

Simonis, A.E., Brownell, R.L., Thayre, B.J., Trickey, J.S., Oleson, E.M., Huntington, R. and Baumann-Pickering, S. 2020. 'Co-occurrence of beaked whale strandings and naval sonar in the Mariana Islands, Western Pacific'. *Proceedings of the Royal Society, B.* doi:10.1098/rspb.2020.0070

Royal Navy. 2020. Joint Warrior Exercises Impact Statement. September 2020.

Hartny-Mills, L. 2021. 'Europe's largest military exercise begins today in the Hebrides'. Hebridean Whale and Dolphin Trust blog. 20 May 2021. https://hwdt.org/news/2021/5/10/europes-largest-military-exercise-begins

AQUARIUS: Folklore and fables

Zumbansen, N. 2018. 'The origin of the Mythical Loch Ness Monster Nessie in the Hagiography of St Columba'. https://freedom-factsandstories.com/2018/ll/06/the-origin-of-the-mythical-loch-ness-monster-nessie-in-the-hagiography-of-st-columba

Adomnán of Iona. *1995 Life of St Columba.* London: Penguin Classics.

Saint Ronan's School. 'Who was St Ronan?' https://www.saintronans.co.uk/history/who-was-st-ronan

Robert Burns. 1785. 'Address to the Deil'. BBC. https://www.bbc.co.uk/arts/robertburns/works/address_to_the_deil

Story of Natsilane in Tlingit culture: https://en.wikipedia.org/wiki/Natsilane

Vancouver Island Map of First Nations. http://viea.ca/business-living-on-vancouver-island/first-nations/

Peterson, L. 2005. 'Tsux'iit: Understanding indigenous spirituality'. *The Dominion, news from the grassroots.* http://www.dommionpaper.ca/original_peoples/2005/05/20/tsuxiit_un.html

The Megalithic Portal. 'Sproat Lake Petroglyphs. Rock art in Canada'. 14 March 2020. https://www.megalithic.co.uk/article.php?sid=37837

Akhlut in Inuit mythology. 2018. 'Mythical creater the akhlut an inuit orca-wolf hybrid with writing prompt'. Arjungwriter.

Bradford, H. 2006. 'Te whānau puha – whales: Whales in Māori tradition', *Te Ara – The Encyclopedia of New Zealand.* https://teara.govt.nz/en/te-whanau-puha-whales/page–1

Kirby, D. 2014. 'Deadly mystery: why did 9 killer whales die in New Zealand?' *Take Part.*

Roy, E. A. 2018. 'Seascape: the state of our oceans. What is the sea telling us? Maori tribes fearful over whale strandings'. *The Guardian.* 3 January 2019. https://www.theguardian. com/environment/2019/jan/03/what-is-the-sea-telling-us-maori-tribes-fearful-over-whale-strandings?fbclid=IwARIdoYLPQISBvSxPqFndWr9hNEPV5kNWnT_961Wj62kv8NxdN-di8mgkj1E

OCCASUS: The sun sets in the west

Weelden, C., Towers, J.R. and Bosker, T. 2021. 'Impacts of climate change on cetacean distribution, habitat and migration'. *Climate Change Ecology* 1. doi.org/10.1016/j.ecochg.2021.100009

Macleod, C.D., Bannon, S.M., Pierce, G., Schweder, C., Learmonth, J.A.,

Hermon, J.S. and Reid, R.J. 2005. 'Climate change and the cetacean community of north-west Scotland'. *Biological Conservation* 124, 477–83.

Piantadosi, C. and Thalmann, E. 2004. 'Whales, sonar and decompression sickness'. *Nature* 428, 1–2. doi:10.1038/nature02527a

WWF. 2020. Living Planet Report, 2020. https://f. hubspotusercontent20.net/ hubfs/4783129/LPR/PDFs/ENGLISH-SUMMARY.pdf

Reeves, R., Pitman, R.L. and Ford, J.K.B. 2017. *Orcinus orca. The IUCN Red List of Threatened Species* 2017, e.T15421A50368125. doi:10.2305/IUCN.UK. 2017–3.RLTS.T15421A50368125.en

Saulitis, E. 2013. *Into Great Silence: A Memoir of Discovery and Loss among Vanishing Orcas.* Boston, MA Beacon Press.

Centre for Biological Diversity. 'A Deadly Toll – The devastating wildlife effect of Deepwater Horizon – and the next catastrophic oil spill'. https://www. biologicaldiversity.org/programs/public_lands/energy/dirty_energy_ development/oil_and_gas/gulf_oil_ spill/a_deadly_toll.html

Hayes, S.A., Josephson, E., Maze-Foley, K. and Rosel, P.E. 2016. US Atlantic and Gulf of Mexico Marine Mammal Stock Assessments 2016. NOAA Technical Memorandum NMFS NE 241.

Esteban, R., Verborgh, P., Gauffier, P., Gimenez, J., Guinet, C. and de Stephanis, R. 2016. 'Dynamics of killer whales, bluefin tuna and human fisheries in the Strait of Gibraltar'. *Biological Conservation* 194, 31–8.

Marine Mammal Protected Area Task Force: Strait of Gibraltar and Gulf of Cadiz, IMMA. https://www.marinemammalhabitat.org/portfolio-item/ strait-gibraltar-gulf-cadiz/

Ford, J.K.B., Pilkington, J.F., Reira, A., Otsuki, M., Gisborne, B., Abernethy, R.M., Stredulinsky, E.H., Towers, J.R. and Ellis, G.M. 2017. 'Habitats of Special Importance to Resident Killer Whales (*Orcinus orca*) off the West Coast of Canada'. DFO Canadian Science Advisory Secretariat. Research document 2017/035.

Bigg, M.A., Olesiuk, P.F., Ellis, G.M., Ford, J.K.B. and Balcomb, K.C. 1990. Social organizations and genealogy of resident killer whales (*Orcinus orca*) in the coastal waters of British Columbia and Washington State. *Report of the International Whaling Commission*, Special Issue 12, 383–405.

Ford, J.K.B. 2006. 'An assessment of critical habitats of resident killer whales in waters off the Pacific Coast of Canada'. Canadian Science Advisory Secretariat Research Document 2006/72.

Ford, J.K.B. 2012. Resident killer whale feeding habits: assessment methods, winter diet, and chum stock ID. Evaluating the Effects of Salmon Fisheries on Southern Resident Killer Whales: Workshop 2, 13–15 March 2012. NOAA Fisheries and Fisheries and Oceans Canada: Vancouver, BC.

Ford, J.K.B. and Ellis, G.M. 2006. 'Selective foraging by fish-eating killer whales *Orcinus orca* in British Columbia'. *Marine Ecology Progress Series* 316, 185–99.

Ford, J.K.B., Ellis, G.M., Barrett-Lennard, L.G., Morton, A.B., Palm, R.S. and Balcomb, K.C. 1998. 'Dietary specialization in two sympatric population of killer whales (*Orcinus orca*) in coastal British Columbia and adjacent waters'. *Canadian Journal of Zoology* 76, 1456–71.

Ford, J.K.B., Ellis, G.M. and Balcomb, K.C. 2000. *Killer Whales: The Natural History and Genealogy of* Orcinus orca *in British Columbia and Washington State*, 2nd edn Vancouver: UBC Press.

Marine Mammal Commission. Southern Resident Killer Whales Population Details. https://www.mmc.gov/priority-topics/species-of-concern/southern-resident-killer-whale/population

The Center for Whale Research. https://www.whaleresearch.com

Desforges, J-P., Hall, A., McConnell, E., Rosing-Asvid, A., Barber, J.L., Brownlow, A., De Guise, S., Eulaers, I., Jepson, P.D., Letcher, R.J., Levin, M., Ross, P.S., Samarra, F., Vikingson, G., Sonne, C. and Dietz, R. 2018. 'Predicting global killer whale population collapse from PCB pollution'. *Science* 361, 1373–76.

MONEYPENNY: Protecting 007

Wasser, S.K., Lundin, J.I., Ayres, K., Seely, T., Giles, D., Balcomb, K., Hempelmann, J., Parsons, K. and Booth, R. 2017. 'Population growth is limited by nutritional impacts on pregnancy success in endangered Southern Resident killer whales (*Orcinus orca*)'. *FLOS one* 12(6), e0179824. doi:10.1371/journal. pone.0179824

Center for Whale Research. https://www.whaleresearch.com/about-orcas

Arsenault, C. 2020. 'Countries fall short of U.N. pledge to protect 10% of the ocean by 2020'. *Mongabay.* 2 December 2020. https://news.mongabay.com /2020/12/countries-fall-short-of-u-n-pledge-to-protect–10-of-the-ocean –by–2020

Scottish Government. 'Protection for harbour porpoise: Europe's largest Special Area of Conservation announced'. 25 September 2016. https:// www.gov.scot/news/protection-for-harbour-porpoise

Hartny-Mills, L. 2020. 'Protected areas for whales, dolphins and sharks designated in Scottish Waters'. Hebridean Whale and Dolphin Trust blog. 3 December 2020. https://hwdt.org/news/2020/12/03-marine-protected-areas-for-whales-dolphins-and-sharks-in-scotland

Beck, S., Foote, A.D., Kotter, S. and Harries, O. 2013. 'Using opportunistic photo-identification to detect a population decline of killer whales (*Orcinus orca*) in British and Irish waters'. *Journal of the Marine Biological Association of the UK*. doi:10.1017/S0025315413001124

Part Two: THE VOYAGE OF THE SILURIAN

Stornoway and the playful minke whale

Bain, C. 2013. *The Ancient Pinewoods of Scotland*. Sheffield: Sandstone Press.